Ultrasonic Transducer Materials

ULTRASONIC TECHNOLOGY
A Series of Monographs

General Editor
Lewis Balamuth
Ultrasonic Systems, Inc., New York, N.Y.

1967: RAYLEIGH AND LAMB WAVES
Physical Theory and Applications
I. A. Viktorov

1969: SOURCES OF HIGH-INTENSITY ULTRASOUND
Volume 1
Edited by L. D. Rozenberg

SOURCES OF HIGH-INTENSITY ULTRASOUND
Volume 2
Edited by L. D. Rozenberg

1971: ULTRASONIC TRANSDUCER MATERIALS
Edited by O. E. Mattiat

HIGH-INTENSITY ULTRASONIC FIELDS
Edited by L. D. Rozenberg

Ultrasonic Transducer Materials

Edited by

O. E. Mattiat

Contributors

Don Belincourt
Vernitron Piezoelectric Division
Bedford, Ohio

Yoshimitsu Kikuchi
Research Institute of Electrical Communication
Tohoku University
Sendai, Japan

A. H. Meitzler
Bell Telephone Laboratories
Murray Hill, New Jersey

Springer Science+Business Media, LLC

ISBN 978-1-4757-0470-9 ISBN 978-1-4757-0468-6 (eBook)
DOI 10.1007/978-1-4757-0468-6

Library of Congress Catalog Card Number 71-131885

SBN 306-30501-1

Originally published by Plenum Press, New York in 1971.

Softcover reprint of the hardcover 1st edition 1971

A Division of Plenum Publishing Corporation
227 West 17th Street, New York, N.Y. 10011

United Kingdom edition published by Plenum Press, London
A Division of Plenum Publishing Company, Ltd.
Davis House (4th Floor), 8 Scrubs Lane, Harlesden, NW10 6SE, England

PREFACE

In recent years remarkable progress has been made in the development of materials for ultrasonic transducers. There is a continuing trend towards increasingly higher frequency ranges for the application of ultrasonic transducers in modern technology. The progress in this area has been especially rapid and articles and papers on the subject are scattered over numerous technical and scientific journals in this country and abroad. Although good books have appeared on ultrasonics in general and ultrasonic transducers in particular in which, for obvious reasons, materials play an important part, no comprehensive treatise is available that represents the state-of-the-art on modern ultrasonic transducer materials. This book intends to fill a need for a thorough review of the subject.

Not all materials are covered of which, theoretically, ultrasonic transducers could be made but those that are or may be of technical importance and which have inherent electroacoustic transducer properties, i.e., materials that are either magnetostrictive, electrostrictive, or piezoelectric. The book has been devided into three parts which somewhat reflect the historic development of ultrasonic transducer materials for important technical application.

Chapter 1 deals with magnetostrictive materials, magnetostrictive metals and their alloys, and magnetostrictive ferrites (polycrystalline ceramics). The metals are useful especially in cases where ruggednes of the transducers are of overriding importance and in the lower ultrasonic frequency range. The fundamentals of magnetostriction, eddy current effects, the relations between static and dynamic effects and methods of measurement precede the description of the magnetostrictive materials themselves. The rationalized mks system has been used throughout but the previously common cgs familiar with the older units.

Both of the following chapters treat materials with electric field effects.

The chapter division here is in the frequency range. Chapter 2 covers piezoelectric crystals as well as "piezoelectric" ceramics. (Strictly speaking the latter should be called "electrostrictive" because the piezoelectric effect is linear whereas the electrostrictive effect is not; but as polycrystalline ceramics can be linearized by prepoling, the term "piezoelectric ceramics" has become usage.) Sections on the fundamental relationships in piezoelectric crystals, important modes of vibration, ferroelectricity and dissipation precede the actual description of the properties of individual piezoelectric materials. Historically, engineering applications started in the lower frequency range (up to a few MHz) and accordingly monolythic crystals and later polycrystalline ceramic materials were developed to meet technical requirements, whereas the microwave frequency range (also called pretersonics) was merely the concern of research workers. Very recently, however, more and more technical interest has developed in the very high ultrasonic frequency range ($>$100 MHz), e.g., in the areas of ultrasonic delay lines, light modulators, and light deflectors, and has made a separate treatment necessary.

Chapter 3 focuses on piezoelectric materials for transducers at pretersonic frequencies. Here the material is intimately linked with special techniques of bonding and lapping, of evaporation and sputtering of films (of the material) onto other substrates. Descriptions of these techniques were, therefore, included in Chapter 3. A number of materials less suitable for lower frequency transducers have become of great importance at pretersonic frequencies and the development of these materials is still proceeding at a rapid pace.

The shift of scientific and technical interest in ultrasonic transducer materials from magnetic field transducers (magnetostrictive materials) to electric field transducers (electrostrictive and piezoelectric materials) and from lower to ever higher frequency ranges is reflected in the fact that roughly one half of the cited literature references in Chapter 1 appeared in the last decade, in Chapter 2 in the last five years, and in Chapter 3 in only the last three years.

May, 1970

OSKAR E. MATTIAT

CONTENTS

Chapter 1

Magnetostrictive Metals and Piezomagnetic Ceramics as Transducer Materials

Yoshimitsu Kikuchi

Chapter 2

Piezoelectric Crystals and Ceramics

Don Berlincourt

Chapter 3

Piezoelectric Transducer Materials and Techniques for Ultrasonic Devices Operating Above 100 MHz

A. H. Meitzler

Chapter 1

MAGNETOSTRICTIVE METALS AND PIEZOMAGNETIC CERAMICS AS TRANSDUCER MATERIALS

Yoshimitsu Kikuchi

Research Institute of Electrical Communication
Tohoku University
Sendai, Japan

1. INTRODUCTION

Magnetostrictive materials have been studied and developed during the last 40 years mainly for the generation and detection of ultrasonic waves. More recently some resonator applications have been introduced. We shall give a few examples of practical magnetostrictive transducers for both generators and detectors as well as electric wave filter applications.

1.1. Ultrasonic Generators and Detectors

Magnetostrictive vibrators are used as electroacoustic transducers to generate or to receive ultrasonic waves. In ultrasonic depth sounders pulsed electrical energy excites the transducer to transmit an ultrasonic wave train into the water towards the bottom from where an echo is reflected to the transducer and converted back into electrical energy. The total travel time of the sound transmission determines the distance (depth). The same principle is applied in ultrasonic fish-finders where the echoes are obtained from schools of fish. In ultrasonic sonar apparatus the direction of ultrasonic transmission and reception is usually horizontal and targets can be detected and measured by their distance and azimuthal direction.

Sonic energy produced by ultrasonic transducers is also directly used in various industrial applications. Ultrasonic cavitation in liquid media causes physical and chemical reactions. Ultrasonic shakers or ultrasonic cleaners of various functions and sizes are now commercially available.

The mechanical vibration of ultrasonic frequency is also industrially utilized in such applications as impact grinding[1] of hard materials, ultrasonic aluminum soldering,[2] ultrasonic impact welding[3,4] of metal pieces, and ultrasonic acceleration of thin-wire drawing.[5] In these applications, a solid horn[2] is generally used to mechanically amplify the vibrational amplitude of the electromechanical transducer.

The frequency of the ultrasonic waves in these applications is usually between 5 and 100 kHz; magnetostrictive transducers are widely used as substitutes for electrostrictive or piezoelectric transducers. The basic shapes of magnetostrictive transducers are shown in Fig. 1; the physical dimensions of a transducer are designed[6,7] so that the whole body (the core of the electrical winding) will be mechanically resonant at the ultrasonic operating frequency.

When the magnetostrictive material is metal, the body is composed of cemented laminations of thin sheets of material so that the eddy-current loss may be kept small. This is a major factor controlling the electroacoustic efficiency of the transducer.[6,7]

More recently, suitable ferrite magnetostrictive materials have been invented and industrially developed. As ferrites generally have very high electrical resistivity eddy-current losses are small even in a bulky lump of the material, and therefore, the shapes shown in Fig. 1 are fabricated by a single molding and sintering of ferrite powder. Ferrite transducers with biasing magnets, which are inserted or bonded to the magnetostrictive parts, have been proposed.[8,9,10] Examples are shown in Fig. 2.

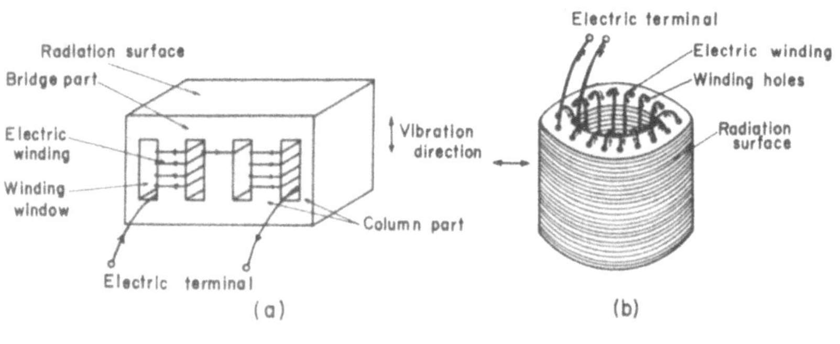

Fig. 1. Basic shapes of magnetostrictive vibrators for ultrasonic use; (a) rectangular type, (b) ring type.

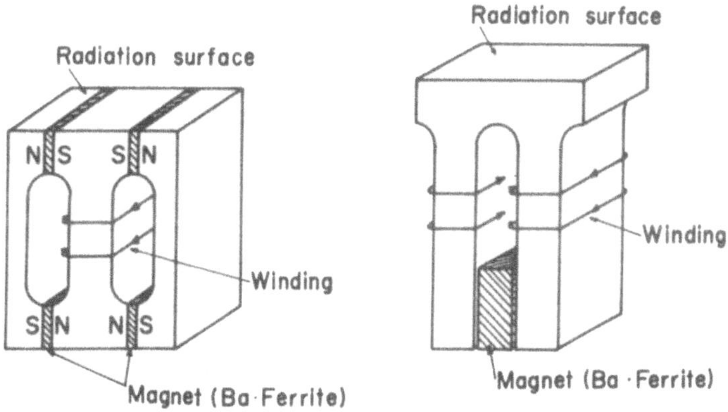

Fig. 2. Magnetostrictive vibrators made of ferrite, biased by permanent magnets (the shaded parts in the figures).

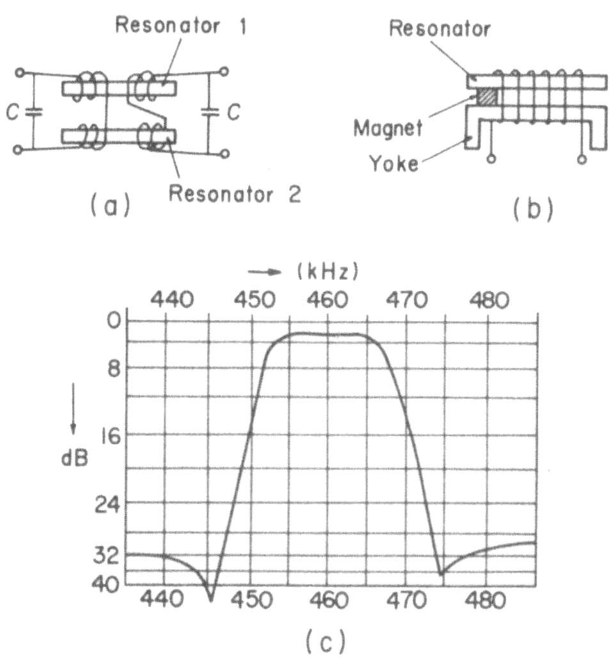

Fig. 3. Magnetostrictive filter of differential type; (a) circuit, (b) ferrite resonator with yoke and magnet, and (c) an example of the measured frequency response.

1.2. Magnetostriction Filters

Magnetostrictive transducers, as components of electric wave filters, were suggested by several authors.[11,12] But, industrial application of these devices was rather limited. In recent years, however, new magnetostrictive resonators made of ferrite have been introduced.[13,14] Magnetostriction filters of the differential type[12,15] were reinvestigated by Shibayama *et al.*[14] using Ni–Cu–Co ferrite (Vibrox-2) as longitudinal resonators. An example of the device and its frequency characteristics are shown in Fig. 3. The magnetostrictive resonators for filter applications must have a small temperature coefficient of resonance frequency and a suitable value of electromechanical coupling factor k together with a high quality factor Q of mechanical resonance. The value of k determines the band width.[16] For narrow band-pass filters the apparent k of the resonator material can be reduced by using a yoke for a magnetic shunt.[16] In this case high-permeability ferrite was used for the yoke material.

Simple twisting resonators were used by Thiele[13] for constructing a narrow-band magnetostriction filter of another type. Boynton[17] reported on a random-wave analyzer which was composed of 80 magnetostriction filters, resonating at about 100 kHz.

2. FUNDAMENTALS OF MAGNETOSTRICTION

2.1. Static Magnetostriction Phenomena

A magnetic substance, in general, changes its physical length when magnetized. This phenomenon is called the Joule effect. Figure 4 shows the Joule effect for several magnetic materials; l is the length of the substance, δl the change in length, and $\delta l/l$ the relative change or "strain," and H the magnetic field.

Observing the individual curves, it is found that some have steep and some have gradual slopes in weak magnetic fields, and that materials of greater saturation magnetostriction do not necessarily have a steep slope and those of smaller saturation magnetostriction do not necessarily have a gradual slope. These facts in major cases are due to the appearance of characteristics concerning the magnetic susceptibility of the materials. If one represents the Joule effect in relation to the intensity of magnetization M instead of H, the appearance of the curves will drastically change. For example, Figs. 5 and 6 show the strain curves for iron and nickel, $\delta l/l$ versus H and M.

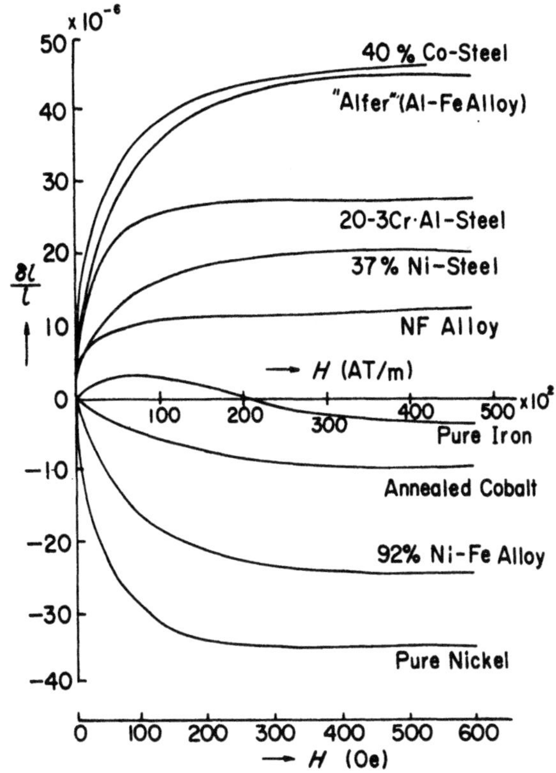

Fig. 4. Static magnetostriction of ferromagnetic materials.

2.2. Magnetostrictive Forces

2.2.1. Magnetostrictive Driving Force

Although the Joule effect produces extremely small strains, very large external mechanical forces will be required to block the strains due to magnetostriction. When an increment of magnetization ΔM is added to a substance which is under a magnetized state M_{DC}, a certain increment of the dimension Δl is yielded. In order to prevent this dimension change, the required external force $\Delta\tau$ is

$$\Delta\tau = -E \cdot \Delta(\delta l/l) \tag{1}$$

where E is the Young's modulus of elasticity and $\Delta(\delta l/l)$ is the corresponding increment of static magnetostriction. In nickel, for example, when $\Delta M = 0.03$ Wb/m² (25 emu), for $M_{DC} = 0.3$ Wb/m² (250 emu), we have

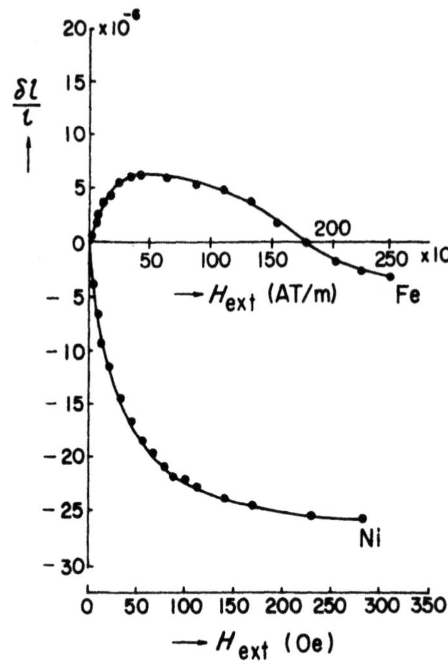

Fig. 5. Static magnetostriction of iron and nickel as a function of external magnetic field.

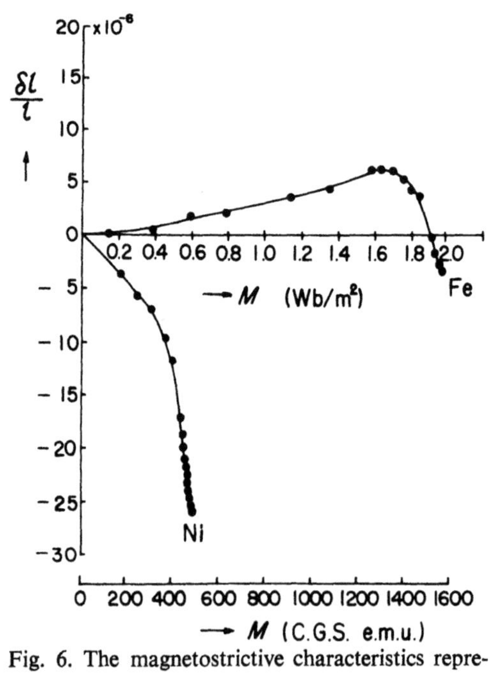

Fig. 6. The magnetostrictive characteristics represented as a function of intensity of magnetization.

$\Delta(\delta l/l) \simeq 0.5 \times 10^{-6}$. And, if we use $E = 2 \times 10^{11}$ Newton/m^2 (2×10^6 kg/cm^2), $\Delta\tau$ will be as large as 10^5 N/m^2 (1 kg/cm^2).

This effect can be interpreted as a force generation in the substance by an amount $\Delta\tau' = -\Delta\tau$ corresponding to the increment of magnetization ΔM. This $\Delta\tau'$ is called the magnetostrictive driving force (or sometimes simply the magnetostrictive stress).

A proportional relation may hold between $\Delta\tau'$ and ΔM if ΔM is sufficiently small. Therefore, we can express the phenomenon as follows:

$$\Delta\tau' = \Gamma \Delta M \tag{2}$$

The coefficient Γ is a magnetostrictive dynamic constant.

2.2.2. Magnetostrictive Magnetomotive Force

The magnetic property of a magnetostrictive material changes when a mechanical strain is applied. This effect can be interpreted as the phenomenon in which a magnetomotive force $\Delta H'$ is induced in the interior of the substance by an increment of mechanical strain $\Delta\xi$, resulting in a corresponding increase of its magnetization. Expressing this phenomenon by an equation, we have

$$\Delta H' = \Gamma' \Delta\xi \tag{3}$$

where Γ' is another magnetostrictive dynamic constant which can be shown to be equal to Γ (*cf.* Section 2.3.2).

2.3. Magnetostrictive Constants

2.3.1. Dynamic Constant

The physical quantities regarding the magnetic state of a matter are the magnetic field H, the intensity of magnetization M, the mechanical stress τ, and the mechanical strain ξ. First, let us consider a function

$$\xi = \xi(\tau, M) \tag{4}$$

where M is a function of H and ξ

$$M = M(H, \xi) \tag{5}$$

The total differential of Eq. (4), the incremental strain $\Delta\xi$, is expressed as[22]

$$\Delta\xi = \frac{1}{E}(\Delta\tau + \Gamma\Delta M) \tag{6}$$

where

$$\Gamma = \left(\frac{\partial \tau'}{\partial M}\right)_\xi \tag{7}$$

and

$$\frac{1}{E} = \left(\frac{\partial \xi}{\partial \tau}\right)_M \tag{8}$$

The dynamic constant Γ thus derived has been called the magnetostrictivity, E being Young's modulus. When the phenomenon changes sinusoidally, various quantities are expressed by complex effective values and we obtain

$$\xi = \frac{1}{E}(\tau + \Gamma M) \tag{9}$$

where $\Gamma M = \tau'$, which is the dynamic expression of Eq. (2). Similarly, we have

$$\Delta M = \varkappa(\Delta H + \Gamma' \Delta\xi) \tag{10}$$

where

$$\Gamma' = \left(\frac{\partial H'}{\partial \xi}\right)_M \tag{11}$$

and

$$\varkappa = \left(\frac{\partial M}{\partial H}\right)_\xi \tag{12}$$

The dynamic constant Γ' is the inverse magnetostrictivity. For the phenomenon of sinusoidal change, using complex effective values, we have

$$M = \varkappa(H + \Gamma'\xi) \tag{13}$$

where $\Gamma'\xi = H'$, which is the dynamic expression of Eq. (3).

2.3.2. Reciprocity Between Γ and Γ'

The reciprocity between the two magnetostrictivities expressed by Eqs. (7) and (11) is easily proved by a method of thermodynamics. However, on the reciprocity between the dynamic magnetostrictivities in alternating phenomena expressed by Eqs. (9) and (13), only the reciprocal relation of their absolute values has been proved[23] by applying an energy conservation principle. But their phase quantities cannot be proved to be reciprocal. Only the fact that they are conjugate quantities has been derived.[23] Namely,

$$|\Gamma| = |\Gamma'| \qquad \text{and} \qquad \angle\Gamma + \angle\Gamma' = 0$$

This relation shows either: (i) $\angle\Gamma = -\angle\Gamma'$ or (ii) $\angle\dot{\Gamma} = \angle\dot{\Gamma}' \equiv 0$. Many experimental results[23] and some considerations[22] through the concept of magnetic domain, however, show that the relation should be similar to (ii). Therefore, we conclude

$$\Gamma = \Gamma' \equiv |\Gamma| \angle 0 \tag{14}$$

2.3.3. Other Definitions of Magnetostriction Constants*

Magnetostriction constants K and K' defined by Smith[24] are

$$\tau' = KH$$

and (15)

$$M' = K'\xi$$

in which τ' is the magnetostrictive driving force, M' is the magnetization generated by the inverse effect (Villari's effect), and ξ is the longitudinal mechanical strain inside the substance.

As the magnetostrictivities Γ and Γ' defined by Kikuchi (see Section 2.3.1 or Ref. (23)) are

$$\tau' = \Gamma M \tag{16a}$$

$$H' = \Gamma'\xi \tag{16b}$$

we have the following correlation in complex effective values

$$K = \varkappa\Gamma \qquad \text{and} \qquad K' = \varkappa\Gamma' \tag{17}$$

though Smith did not touch upon the phase angle due to eddy current and hysteresis. If we introduce into these relations the relation of $\angle\Gamma = \angle\Gamma' \equiv 0$, it will be easily seen that the following relations exist

$$\angle K = \angle K' \equiv \angle\varkappa \tag{18}$$

Later, Butterworth and Smith[25] defined the magnetostrictive constants λ and γ in the following form:

$$\begin{aligned} \tau' &= \lambda B \\ H' &= \gamma\xi \end{aligned} \tag{19}$$

* Though all of the constants in this section have originally been defined in CGS electromagnetic units, we first describe them in rationalized units, and then show the cross-reference table in both units.

Table I. Cross-Reference of Magnetostriction Constants

Notation	K, K' (Smith/Aoyagi)	λ, γ (Butterworth/Smith)	Γ, Γ' (Kikuchi)
mks rationalized units			
Γ, Γ'	$K = \varkappa\Gamma$ $K' = \varkappa\Gamma'$	[a] $\lambda = \Gamma$ $\gamma = \Gamma'$	—
λ, γ	$K = \varkappa\lambda$ $K' = \varkappa\gamma$	—	[a] $\Gamma = \lambda$ $\Gamma' = \gamma$
K, K'	—	[a] $\lambda = K/\varkappa$ $\gamma = K'/\varkappa$	$\Gamma = K/\varkappa$ $\Gamma' = K'/\varkappa$
cgs electromagnetic units [c]			
Γ, Γ'	$K = \varkappa\Gamma$ $K' = \varkappa\Gamma'$	[b] $\lambda = \Gamma/4\pi$ $\gamma = \Gamma'$	—
λ, γ	$K = 4\pi\varkappa\lambda$ $K' = \varkappa\gamma$	—	[b] $\Gamma = 4\pi\lambda$ $\Gamma' = \gamma$
K, K'	—	[b] $\lambda = K/4\pi\varkappa$ $\gamma = K'/\varkappa$	$\Gamma = K/\varkappa$ $\Gamma' = K'/\varkappa$

[a] Provided that $B \simeq M$.
[b] Provided that $B \simeq 4\pi M$.
[c] See Table II (p. 26) for conversion ratios.

where B is the magnetic induction. So, directly from Eq. (16) we have

$$\gamma = \Gamma' \tag{20}$$

As to the constant λ, however, assuming $B \simeq M$ in the relation between magnetic induction B and intensity of magnetization M, we have

$$\lambda = \Gamma \tag{21}$$

2.3.4. Cross-Reference between K, λ, γ, Γ, and Γ'

The three pairs of Eqs. (15), (16), and (19) will disclose the cross-reference of the three systems of dynamic magnetostrictive constants. The result is shown in Table I* in which $\varkappa$ represents the magnetic susceptibility (a complex number) at a certain frequency ω, expressing the vector value of the ratio of the alternating M and H, or M' and H' at a displaced (bias)

* K (which is equal to $\varkappa\lambda$ or $\mu\lambda$), is sometimes called the relative open-circuit sensitivity.[26]

dc magnetization. In each system of the magnetostrictive constants, reciprocal relations $K = K'$, $\gamma = \lambda$, or $\Gamma = \Gamma'$ can be proved by the energy conservation principle.

2.3.5. Electromechanical Coupling Factor*

The electromechanical coupling factor is the square root of the ratio of the mechanical energy W_m in a ferromagnetic substance stored by the magnetostrictive effect and the electrical energy W_e stored electrically.† If the phenomena are sinusoidal in character, one can use their respective peak values. When the frequency is far below the mechanical resonance frequency, the mechanical strain in the substance is only ξ' (effective value) which is the usual magnetostriction ($\delta l/l$). So the mechanical energy W_m per unit volume is

$$W_m = \frac{1}{2} E(\sqrt{2}\,\xi')^2 \tag{22}$$

where E is the elastic modulus. With Eqs. (1) and (16a), the following relation holds in a magnetostrictive substance

$$\xi' = \frac{\tau'}{E} = \frac{\Gamma M}{E}$$

Hence,

$$W_m = \frac{\Gamma^2 M^2}{E} \tag{23}$$

By denoting the effective values of the magnetic field and the intensity of magnetization H and M respectively, the magnetic energy W_e sent into the magnetostrictive substance electrically becomes

$$W_e = \frac{1}{2}(\sqrt{2}\,M)(\sqrt{2}\,H) = M \cdot \frac{M}{\varkappa} \tag{24}$$

where $\varkappa = M/H$. Now, the ratio of W_m and W_e becomes

$$\frac{W_m}{W_e} = \frac{\Gamma^2 M^2}{E} \cdot \frac{\varkappa}{M^2} = \frac{\varkappa \Gamma^2}{E} \tag{25}$$

Therefore, the electromechanical or piezomagnetic coupling factor k of

* Sometimes called piezomagnetic coupling factor.

† Magnetic energy is considered to belong to the electrical system in the usage of ultrasonic transducer theory.

the substance is

$$k = \left(\frac{W_m}{W_e}\right)^{1/2} = \left(\frac{\varkappa \Gamma^2}{E}\right)^{1/2} \tag{26}$$

or

$$k = \left(\frac{\mu \lambda \gamma}{E}\right)^{1/2} \tag{27}$$

2.4. Material Criteria

2.4.1. Transducer Efficiency

The electroacoustic transduction efficiency of a magnetostriction vibrator is dependent not only on the material of which the vibrator is made, but also on various factors such as the size and dimensions of the vibrator, the space factor of the laminated material in case where the material consists of metallic sheets, etc. Though the performance characteristics and the design procedure have already been established,(6) separate calculations are needed in individual cases of material selection for actual vibrators; there is no simple relation between the material and the electroacoustic efficiency of the vibrators.

A successful approach for a solution is the concept of potential efficiency.(27,28) For resonators in water, Kikuchi *et al.* proposed some universal charts,(6,29,30,31,9,32) and an example is shown in Fig. 7. These charts are the results of theoretical analyses of rectangular-type vibrators with certain standardizations of their shapes and working conditions. The vibrators

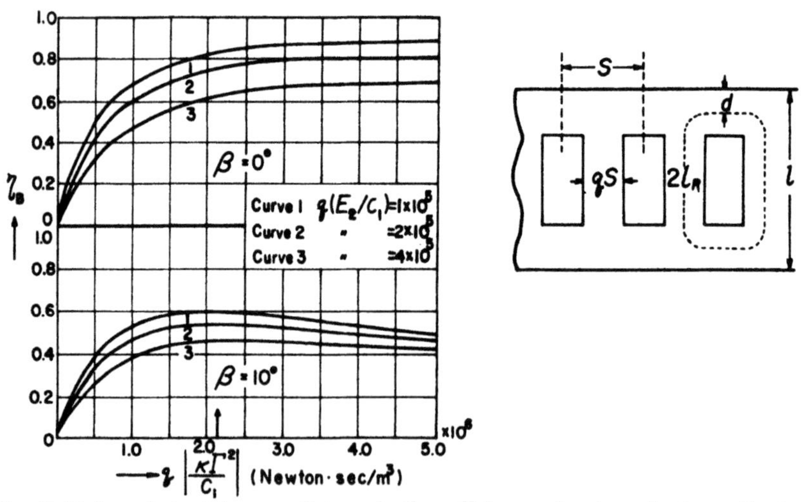

Fig. 7. Universal chart for overall transduction efficiency of rectangular-type vibrators.

radiation surfaces (see Fig. 1a) should be sufficiently large so that the radiation resistance can be assumed purely resistive (equal to the specific acoustic resistance of water times the area of the radiation surface). All other surfaces should be loaded by air. Some dimensional relationships for a vibrator (side view) are shown in Fig. 7; d is neither extremely thin nor thick and $2l_R$ (the length of a closed magnetic circuit) is approximately equal to $2l$, where l is the length of the vibrator, which determines the mechanical resonance condition of the fundamental longitudinal mode,[6] i.e., the operating frequency of the vibrator in the liquid.

The universal chart of overall efficiency (when used at the frequency for the highest transduction efficiency[31]), η_B, of any vibrator can be used in a direct connection to material constants regardless of its resonance frequency. The material constants are $\varkappa\Gamma^2$, c_1 (propagation velocity of longitudinal elastic wave in the material), E_2 (loss component of Young's modulus defined by $E = E_1 - jE_2$), and β (magnetic lagging angle due to eddy current); q is the only factor which relates to the shape of the vibrator.

One can conclude from the chart that it is not necessary to pursue materials having an electromechanical coupling factor greater than values near the point marked by an arrow on the abscissa, if the materials are used for a vibrator working at resonance in liquid.

2.4.2. ΔE-Effect as a Material Criterion

The Young's modulus of a magnetic substance changes with the variation of magnetization; this is called the ΔE-effect. In a substance of isotropic magnetostriction, the total change ΔE from the demagnetized state to the saturation can be expressed[33] as

$$\frac{\Delta E}{E} = C\varkappa_a E\lambda_s^2/M_s^2 \tag{28}$$

where $\varkappa_a$ is the initial permeability, λ_s the saturation magnetostriction, and C a constant nearly equal to unity. On the other hand, Kikuchi and Shimizu,[34,35] using an empirical formula for static magnetostriction (see Section 6.1.5) and Gans' relation on incremental magnetic susceptibility,[36] have derived a formula for the electromechanical coupling factor $k_{\max}$ at optimum bias

$$k_{\max}^2 = 0.565\varkappa_a E\lambda_s^2/M_s^2 \tag{29}$$

From the above two equations, the relation

$$\sqrt{\Delta E/E} = 1.33k_{\max} \tag{30}$$

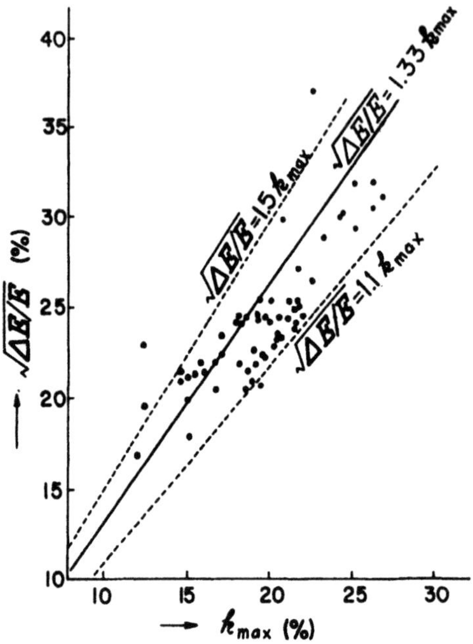

Fig. 8. Experimental relation between ΔE-effect and electromechanical coupling factor at optimum bias k_{max}.

is obtained. Experimental data for various kinds of magnetostrictive ferrite are plotted in Fig. 8, and show fair agreement with the theoretical results of Eq. (30).

2.5. Effect of Hydrostatic Pressure on Magnetostriction

The effect of hydrostatic pressure on dynamic magnetostrictive constants has not yet been fully determined. Karplus[37] observed in 1963 that the output of a laminated nickel-bar 32-kHz vibrator increased monotonically in the range of 1 to 150 atm of the hydrostatic pressure. Kikuchi *et al.*[38] phenomenologically discussed in 1964 the effect on the magnetostrictivity Γ. They concluded that as long as the volume magnetostriction of a substance is negligibly small compared to the longitudinal magnetostriction, no appreciable effect may appear by the impression of hydrostatic pressure. In magnetic substances of isotropic magnetostriction, such as nickel or ferrite, the volume magnetostriction is generally two orders of magnitude smaller than the longitudinal magnetostriction. The pressure independency of $\varkappa$, Γ and E has been confirmed with a magnetostrictive vibrator made of ferrite through experimental observations up to 70 atm in water.[39]

2.6. Effect of Mechanical Stress on Magnetostriction

In various phases of research the following points have been discussed using magnetic-domain theories and have been verified experimentally:[40] (i) In an isotropic magnetostrictive substance, a saturation magnetostriction ξ_s' is equal to the spontaneous deformation λ_s of an atomic lattice when the substance is free from mechanical stress. But ξ_s' decreases, approaching zero, when stressed with the same sign as that of λ_s (i.e., tensile stress when λ_s is positive and compressive stress when λ_s is negative) and ξ_s' increases, approaching $\frac{3}{2}\lambda_s$, when stressed with the opposite sign as that of λ_s. (ii) In the latter case (i.e., opposite sign stress), the static magnetostriction curve $\xi'(M)$ approaches a curve proportional to M^2.

Shimizu *et al.*[41,42] recently reported formulated expressions for magnetization and magnetostriction characteristics in the case where a large stress was applied to a magnetic substance. The results indicate that the expressions are sufficiently applicable over a wide range of stress. One of the important results will be shown here.

The magnetostriction, $\delta l/l \equiv \xi'$, can be expressed as

$$\xi' = \frac{3}{2}\lambda_s h\left(\frac{M}{M_s}\right)^{2/h} \tag{31}$$

where h represents the effect of the stress τ,

$$h = \frac{1}{1 + \frac{1}{2}e^{2\tau/\sigma}} \tag{32}$$

and σ is a sort of material constant having the same sign as that of λ_s.

Since

$$\Gamma = E\left(\frac{\partial \xi'}{\partial M}\right)_\tau \tag{33}$$

(*cf.* 2.3.1), we have

$$\Gamma = 3E\frac{\lambda_s}{M_s}\left(\frac{M}{M_s}\right)^{(2/h-1)} \tag{34}$$

Figure 9 shows the characteristics theoretically derived for the normalized magnetostrictivity $(\Gamma/(3E\lambda_s/M_s))$ as a function of M/M_s, the parameter of the curves being h.

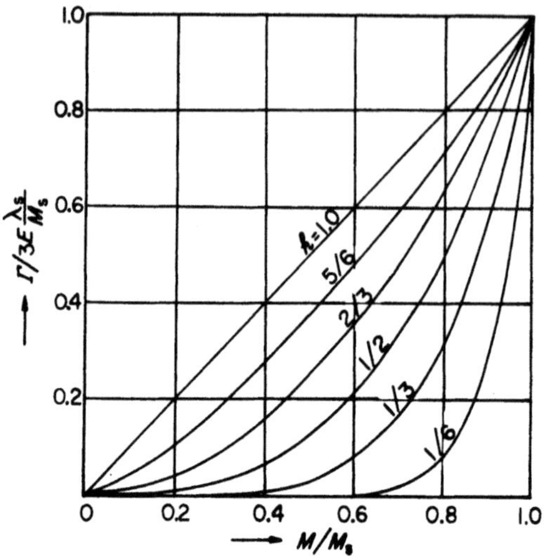

Fig. 9. Theoretically derived characteristics of magnetostrictivity Γ as a function of magnetization.

As seen from Eq. (32), when $\tau = -\infty$, we have $h = 1$. Therefore, if we measure the saturation magnetostriction ξ_s' in the case where the stress is large enough and of opposite sign, the value of h can be obtained from Eq. (31)

$$h = \xi_s'/\tfrac{3}{2}\lambda_s \tag{35}$$

Experimental verification has not yet been available though some related data were disclosed; see Fig. 18 for nickel and Fig. 26 for permendur.

3. EDDY-CURRENT EFFECTS ON MATERIAL CONSTANTS

In magnetizing a metal sheet with an alternating magnetic field, both the intensity of the magnetic field and the intensity of magnetization are nonuniform over the cross section of a sheet because of eddy-current skin effects.

If the mean value of magnetization over a cross section is $\bar{M}$ and the intensity of the magnetic field on the surface of a sheet is H_0, we have[22]

$$\bar{M} = \varkappa(0)H_0\frac{\tanh ma}{ma} \tag{36}$$

where $\varkappa(0)$ is the complex susceptibility with no eddy-current effect, $2a$

is the thickness and m is a skin effect parameter

$$m = \pm(j\varkappa(0)\omega/\sigma)^{1/2} \tag{37}$$

where σ is the specific electrical resistance and $\omega = 2\pi f$ ($f =$ frequency). For later use, we define an apparent susceptibility $\varkappa$,

$$\varkappa = \frac{\bar{M}}{H_0} = \varkappa(0)\frac{\tanh ma}{ma} = \varkappa(0)\cdot\chi \tag{38}$$

If we now take $\bar{M}$ instead of M in Eq. (16a), we have

$$\tau' = \Gamma\bar{M} \tag{39}$$

where $\Gamma\bar{M}$ represents the mean magnetostrictive force over the cross section. Considering the elongation of a whole body, (as long as the size of its cross section is not extremely large), it is correct to assume that the mean value $\Gamma\bar{M}$ operates uniformly over the cross section. Therefore, the relation

$$\xi = \frac{1}{E}(\tau + \Gamma\bar{M}) \tag{40}$$

holds.

Next, consider the eddy-current skin effect on Villari's magnetization. In the case where there is a uniform alternating strain on the cross section of a sheet, the magnetostrictive magnetomotive force, $H' = \Gamma'\xi$, yields a magnetization M' which really exists with a certain distribution over a cross section. By a similar theoretical treatment,[22] the mean magnetization $\bar{M}'$ over the cross section can be expressed as

$$\bar{M}' = M'(a)\frac{\tanh ma}{ma}$$

where $M'(a)$ is the Villari's magnetization at the surface. It is obvious that

$$M'(a) = \varkappa(0)H'$$

Therefore, from Eqs. (36) and (38) we have

$$\frac{\bar{M}'}{H'} = \varkappa(0)\frac{\tanh ma}{ma} = \varkappa \tag{41}$$

and thus we have

$$\bar{M} = \varkappa(H_0 + \Gamma'\xi) \tag{42}$$

where H_0 is the same as in Eq. (36).

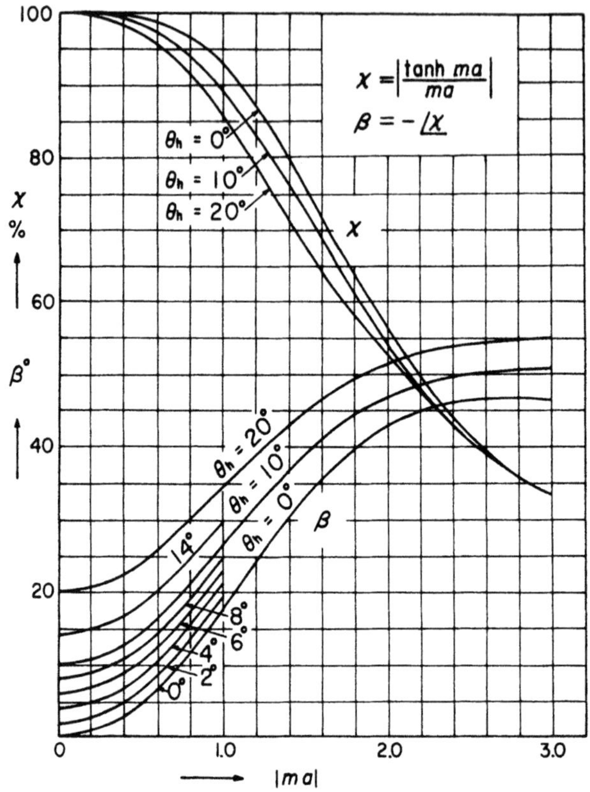

Fig. 10. Eddy-current skin-effect correction for magnetic susceptibility $\varkappa$ (see Eq. (38)); $2a$, thickness of a sheet metal in meter; m, parameter in mks (see Eq. (37)); θ_h, hysteresis angle (see Eq. (45)).

Thus, it becomes clear that magnetostrictive vibration in the case where there is an eddy current can be treated with the fundamental equations in quite the same form as Eqs. (9) and (13).

With this theoretical background, a convenient chart can be drawn. Equation (38) is illustrated in Fig. 10.* In relation to Eq. (38), the corresponding apparent permeability μ is newly defined here as follows

$$\mu = \mu_0 + \varkappa \tag{43}$$

where μ_0 is the permeability of vacuum. For $\mu_0 \ll \varkappa$, we have

$$\angle\mu \simeq \angle\varkappa \tag{44}$$

* This chart was prepared by K. Fukushima.

and also

$$\angle\mu(0) \simeq \angle\varkappa(0) \tag{45}$$

which is the "hysteresis angle" θ_h in Fig. 10.

4. RELATION BETWEEN STATIC AND DYNAMIC MAGNETOSTRICTION PHENOMENA

From the total differential equation of Eq. (4), we can derive an equation

$$\Gamma = \left(\frac{\partial \xi}{\partial H}\right)_\tau \left(\frac{\partial \tau}{\partial \xi}\right)_M \Big/ \left(\frac{\partial M}{\partial H}\right)_\tau \tag{46}$$

where the first term, $(\partial\xi/\partial H)_\tau$, is the slope of a static magnetostriction characteristic curve represented in $\delta l/l$ versus H, observed under constant tensile stress, the second term is the elastic modulus (hereafter represented by E_M) under constant magnetization, and the third term is the slope of an ordinary magnetization curve observed under constant tensile stress, a quantity defined as differential susceptibility (hereafter represented by $\varkappa_\tau$). Equation (46) then becomes $\Gamma = (\partial\xi/\partial H)_\tau \cdot E_M/\varkappa_\tau$. Thus the characteristics of Γ as a function of the bias field H_{DC} can be obtained, but we should remember that there is a hysteresis associated with H_{DC}.

It has been pointed out[23] phenomenologically that a static magnetostriction curve represented in $\delta l/l$ versus M_{DC} has nearly no hysteresis on M_{DC} and is hardly sensitive to internal stress. This means that the phenomenon is a univalent function of M. The slope of this curve, $(\partial\xi/\partial M)_\tau$, can be applied in the relation

$$\Gamma = \left(\frac{\partial \xi}{\partial M}\right)_\tau \left(\frac{\partial \tau}{\partial \xi}\right)_M \tag{47}$$

which is a simple modification of Eq. (46). The solid line in Fig. 11 is obtained from the static characteristics of a pure nickel rod.[22] The circles in Fig. 11 are values obtained from dynamic measurements with a magnetostrictive vibrator made of nickel sheet. The agreement is good.

Conversely, a static magnetostriction curve $(\delta l/l)$ can be obtained from the dynamic characteristics of a magnetostrictive vibrator. As $(\delta l/l)$ is expressed by the integration of $(\partial\xi/\partial M)_\tau$ from $M = 0$ to M, using Eq. (47), we have

$$\frac{\delta l}{l} = \int_0^M \Gamma/E_M \, dM \tag{48}$$

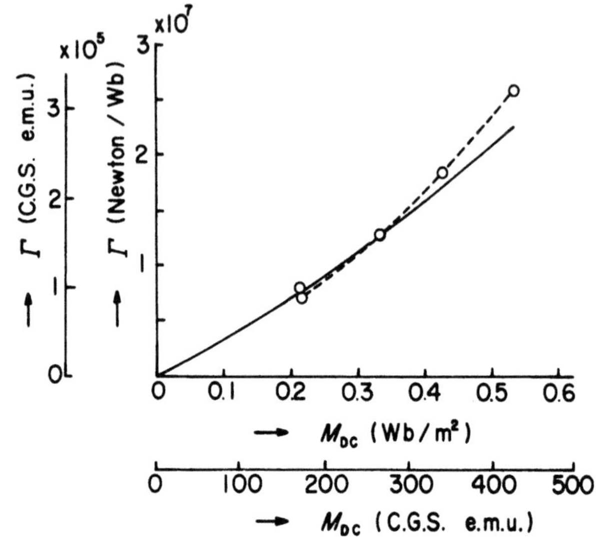

Fig. 11. Experimental verification of Γ derived from static characteristics ($\delta l/l$) in nickel.

The solution is obtained by graphical integration, using the values of Γ and E_M dynamically observed at various M_{DC}.

As described above, we can easily obtain the quantitative relations between the static and alternating phenomena when treated in relation to magnetization; but it is impossible to obtain Smith's constant K from the static curve directly. This is because, though $K = \varkappa_\Delta \Gamma$, the incremental susceptibility $\varkappa_\Delta$ in most cases is very different from the differential susceptibility $\varkappa_\tau$ which is used in Eq. (46). Therefore,

$$K = \varkappa_\Delta \Gamma \neq \varkappa_\tau \Gamma \tag{49}$$

In most cases, $\varkappa_\Delta \Gamma$ is considerably smaller than $\varkappa_\tau \Gamma$.

Using relations described above, the basic characteristics of magnetostriction in pure iron and pure nickel are summarized qualitatively in Fig. 12.

5. METHODS OF MATERIAL MEASUREMENT

5.1. Motional Impedance Methods

Magnetostrictive material is characterized by its magnetic susceptibility $\varkappa$, magnetostrictive constant Γ, γ, or K, Young's modulus E or sound propagation velocity c, and electromechanical coupling factor k, which

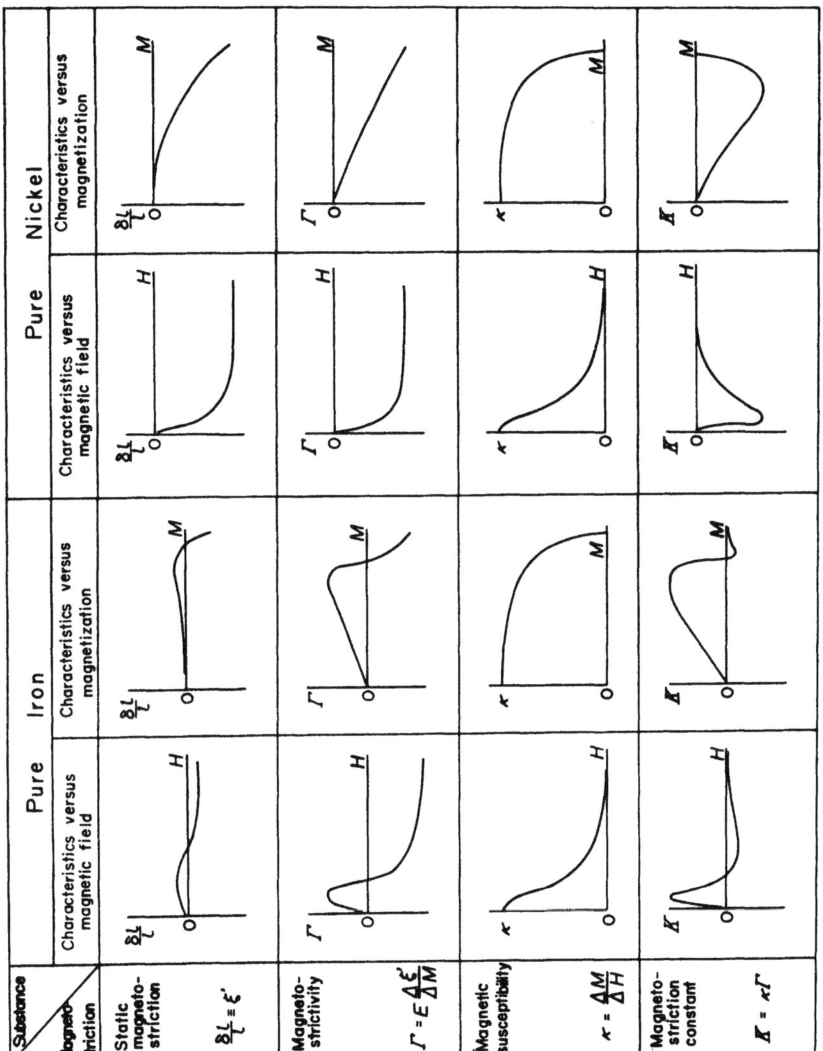

Fig. 12. Characteristics of $\delta l/l$, Γ, $\varkappa$ and K as functions of H_{DC} and M_{DC} in pure iron and pure nickel.

can be expressed by the above quantities, $k = (\varkappa\Gamma^2/E)^{1/2}$. All these parameters are a function of the bias magnetic field or bias magnetization, and determine the performance of a magnetostrictive vibrator.

For small signals, there are well-known measuring techniques using self-motional impedance analysis obtained with an impedance bridge(43,44) on a ring-form specimen or on a slender-strip specimen.(45) Detailed discussions have been given on the dynamic Villari's magnetization in the slender-strip vibrator. For the latter specimen, the demagnetizing effect due to magnetic free poles must be correctly compensated. Kikuchi *et al.*(45) have recently reported the demagnetizing factor $\mathfrak{N}$ for a slender strip of any demension. The numerical table and a nomograph are available in Ref. (45).

A mutual motional impedance method was proposed in 1942(46) to save measurement time, and improved in 1948(47) and 1954,(48) increasing the accuracy of measurement to values almost comparable to the ring-form specimen method. The details are also described in Refs. (45) and (48). The method is convenient when the number of specimens is large.

A vibrometer method and a differential method, which will be described later in Section 7 with regard to large signal operation of vibrators, can also be used for small signal measurement.

In connection with the motional impedance method, an accurate measurement of quadrantal frequency differences has been proposed by Kikuchi *et al.* in 1955. The measuring ac is modulated by a signal of low frequency which is equal to the frequency difference.(45,49)

In the following, some special devices for measuring characteristics under hydrostatic pressure and also under mechanical stress will be described.

5.2. Measurement under Hydrostatic Pressure

Figure 13 is a schematic view of a device reported by Kikuchi *et al.*(50) in 1964 for measuring the performance of a magnetostrictive vibrator under hydrostatic pressure. In a thick metallic pipe containing water a wooden

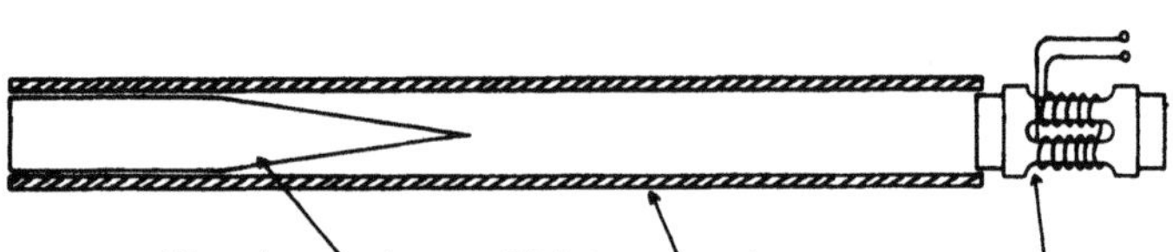

Fig. 13. Acoustic tube for measuring the effect of hydrostatic pressure on magnetostriction characteristics.

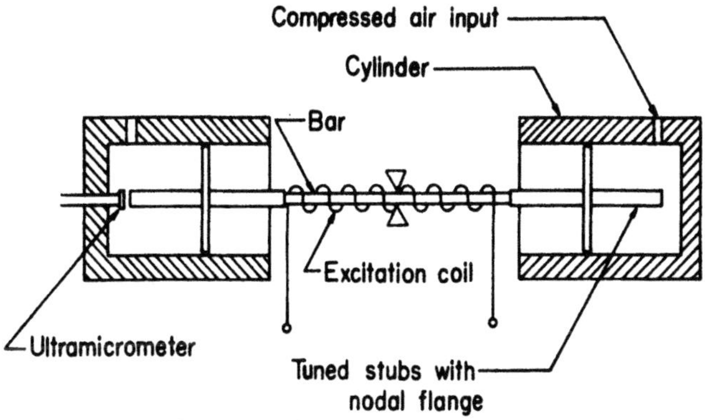

Fig. 14. Mechanical assembly for measuring the effect of mechanical stress on magnetostriction characteristics (after Whymark).

wedge is placed at one end and the magnetostrictive vibrator under observation, with a bonded cylindrical block on its radiation surface, is fitted into the pipe at the other end. The whole system is placed in a pressured water tank. The wedge absorbs the sound waves. A water film at the interface between the pipe and the cylindrical block of the vibrator acts as "acoustic packing." When the pipe is sufficiently thick so that the condition of an acoustic tube may be fulfilled, progressive plane waves can be generated by the magnetostrictive vibrator under a hydrostatic pressure. An acoustic reflector covers the rear surface of the vibrator in order to prevent radiation.

By using a wedge made of pine for the absorber and a stepped solid-horn for the reflector, measurements up to pressures of 70 atm were obtained. The pine wood can be "aged" for its acoustic stability by cycling the hydrostatic pressure several times. Phenomena of ultrasonic cavitation under large hydrostatic pressure[51] can also be observed by this device.

5.3. Measurements under Static Compressive Stress

Figure 14 shows the schematic view of a mechanical assembly reported by Whymark[52] in 1961 for observing the effect of compressive stress on the dynamic magnetostriction behavior of substances. An axial stress is applied to a bar shaped specimen through a pair of half-wavelength rods bonded to the end surfaces of the specimen. At the center of the tuned rod a thin flange is provided to fit into a cylinder and compressed air is impressed on the flange, which acts as a piston. The knife edge located very close to the center of the specimen is merely for the prevention of mechanical buckling of the specimen in case too much stress is applied.

When the ratio of the area of the flange and the cross section of the specimen was on the order of 100 : 1, a stress up to 24×10^6 N/m^2 (240 kg/cm^2) did not affect the mechanical damping of the system. Results on magnetostriction of nickel and 2V-permendur (10 cm long) were reported (see Sections 6.1.2 and 6.4).

6. MAGNETOSTRICTIVE PROPERTIES OF MATERIALS

6.1. Nickel

6.1.1. Static Magnetostriction of Nickel

The static magnetostriction curves ($\delta l/l$ vs H)[53] of a certain pure nickel, observed in the temperature range between 17° and 170°C, showed that the higher the temperature, the steeper the slope of the curve in a weak magnetic field and the smaller the saturation (see Fig. 15).

With another pure nickel, the temperature dependence of its saturation magnetostriction $(\delta l/l)_s$ was observed[53] (see Fig. 16), and it was found that the proportionality constant C in $(\delta l/l)_s = CM_s^2$ is almost invariant with temperature. But the value of C itself cannot be regarded as material constant because the values of $(\delta l/l)_s$ reported by many observers up to 1933 differ from each other, ranging from 22.5×10^{-6} to 46.8×10^{-6}.

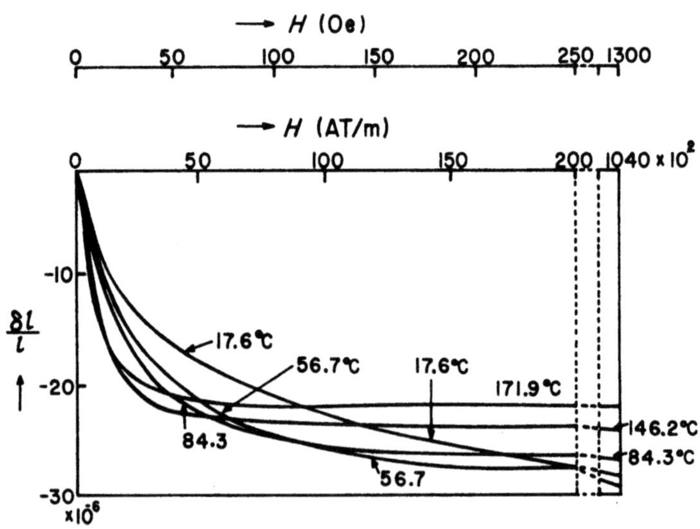

Fig. 15. Temperature dependence of $\delta l/l$ vs H in nickel (after Döring).

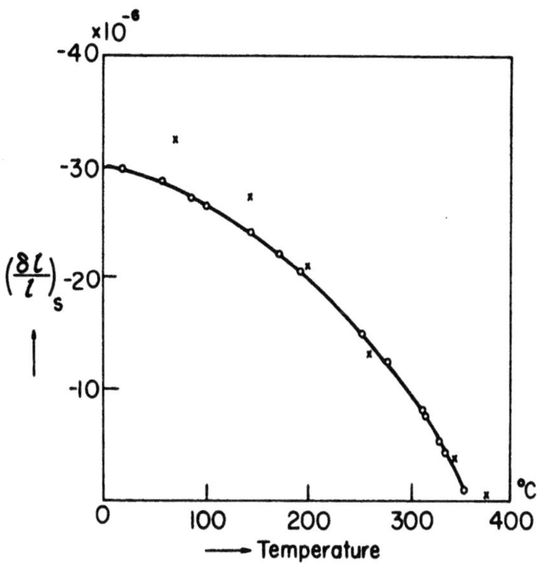

Fig. 16. Temperature dependence of saturation magnetostriction (after Döring).

Static magnetostriction characteristics of nickel under mechanical tension and compression were observed by Kirchner[54] in 1936, and are shown in Fig. 17. The curves $\Delta l/l$ versus M^2 tend to be a straight line when the tension is increased to 104×10^6 N/m² (1040 kg/cm²). The saturation magnetostriction increases from 40×10^{-6} to 60×10^{-6} with the increase of tension from 0 to 120×10^6 N/m² (1200 kg/cm²). But, it decreases, from 40×10^{-6} down to 2×10^{-6}, with the compression of the same order.

6.1.2. Dynamic Magnetostriction of Nickel

Typical characteristics of dynamic magnetostriction in annealed commercial pure nickel sheet are shown in Figs. 11 and 12 in Section 4. The optimum bias for $\varkappa(0)\Gamma$ is around 8×10^2 At/m* (10 Oe), whereas that for $\varkappa(0)\Gamma^2$ is around 20×10^2 At/m (25 Oe). Depending on the source of commercial material and on annealing temperatures, the characteristics may change to some extent. Major data on magnetostrictive metals and ceramics are presented in Table III (see p. 44).

The dynamic characteristics of nickel are rather insensitive to hydrostatic pressure (*cf.* Section 2.5), but are sensitive to tensile and compressive stresses. Theoretical characteristics are already described in Section 2.6.

* At, Ampere-turn.

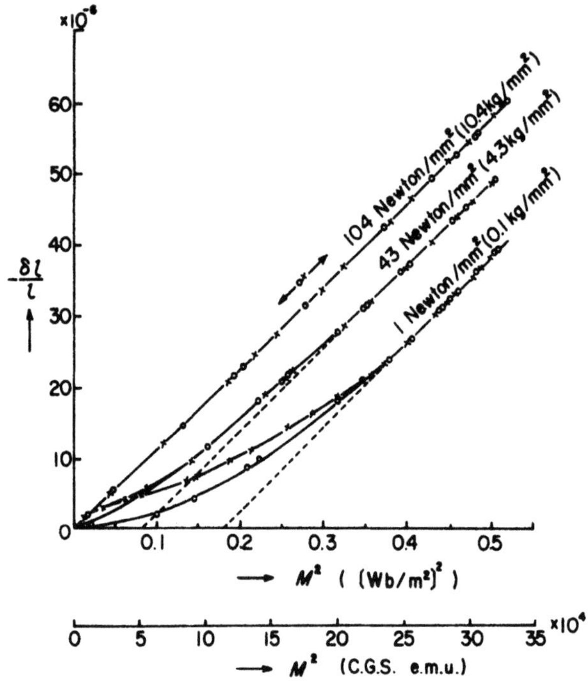

Fig. 17. Static magnetostriction characteristics under mechanical tension, abscissa being M^2 (after Kirchner).

Table II. Conversion Ratios

Notation	cgs emu		mks	
	unit	emu/mks	unit	mks/emu
Magnetostriction constants				
Γ	dyn/cm² · M	$10^3/4\pi$	N/Wb	$4\pi/10^3$
Γ'	Oe	$10^3/4\pi$	At/m	$4\pi/10^3$
λ	dyn/cm² · G	10^3	N/Wb	$1/10^3$
γ	Oe	$10^3/4\pi$	At/m	$4\pi/10^3$
K	dyn/cm² · Oe	$4\pi/10^4$	N/At · m	$10^4/4\pi$
K'	M	$4\pi/10^4$	Wb/m²	$10^4/4\pi$
Susceptibility				
$\varkappa$	emu (M/Oe)	$(4\pi)^2/10^7$	H/m	$10^7/(4\pi)^2$

Unit abbreviations: At, ampere-turns; cm, centimeter; dyn, dyne; G, gauss; H, henry; M, intensity of magnetization in cgs emu; m, meter; N, newton; Oe, oersted; and Wb, weber.

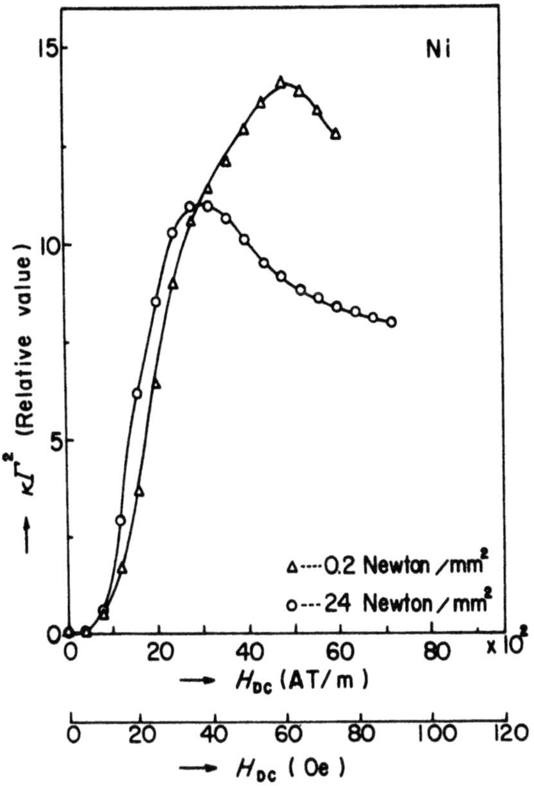

Fig. 18. Relative electromechanical coupling factor squared ($\propto \varkappa\Gamma^2$) in nickel under mechanical compressive stress (after Whymark).

Observed data are rare. Figure 18 shows Whymark's data[52] on the relative electromechanical coupling factor of nickel sheets, which are 0.05 mm (2-mil) thick and have been annealed in air at about 800°C for approximately 2 h. When the compressive stress is raised up to 24 N/mm² (240 kg/cm²), a considerable decrease occurs in its maximum value. The method of the measurement has been described in Section 5.3.

6.1.3. Effect of Cold Reduction

In nickel sheet, the effect of cold working on dynamical magnetostrictive constants and sound velocity was observed[55,45] in 1942 by Kikuchi *et al.* in the range between 20% and 80% of cold-roll reduction in relation with annealing temperatures. Appreciable anisotropy in the characteristics of $\varkappa\Gamma^2$ versus H_{bias} was found with regard to the direction of rolling where the cold work was less than 40% and annealing temperature was sufficiently high (1000°C); the propagation velocity c of longitudinal waves in the samples, however, did not show high anisotropy. But, when the annealing temperature is low (600°C), however, the samples of higher cold work show considerable anisotropy, as high as 4%, in the propagation velocity of longitudinal wave, which is the largest in the transversal direction (90°) and smallest at 45° to the rolling direction. Nevertheless, the characteristics shown of Γ versus M_{bias} were generally insensitive to cold working as well as to annealing temperature in a wide range.[45,56] Theoretical analysis for various textures in nickel sheet was developed by Jones, Clark, *et al.*[57] in 1958 with regard to Young's modulus, magnetostrictive coupling factor, and initial and maximum permeabilities. Some experiments with highly pure polycrystalline nickel sheet, produced by powder metallurgy and annealed in hydrogen, showed fair agreement with the theory.

6.1.4. Hysteresis Phenomena in the Magnetostriction of Nickel

The hysteresis phenomenon is clearly observable[58] in a static magnetostriction curve plotting $\delta l/l$ versus H_{eff}, but it hardly shows up when the same data are plotted versus M. A similar behavior was observed[59] in Γ characteristics as shown in Fig. 19. This was also found for various kinds of sheet material. A useful engineering hypothesis[59] has been set forth, stating that the value of Γ is a unique function of biasing magnetization and independent of the hysteresis between the biasing magnetic field and magnetization.

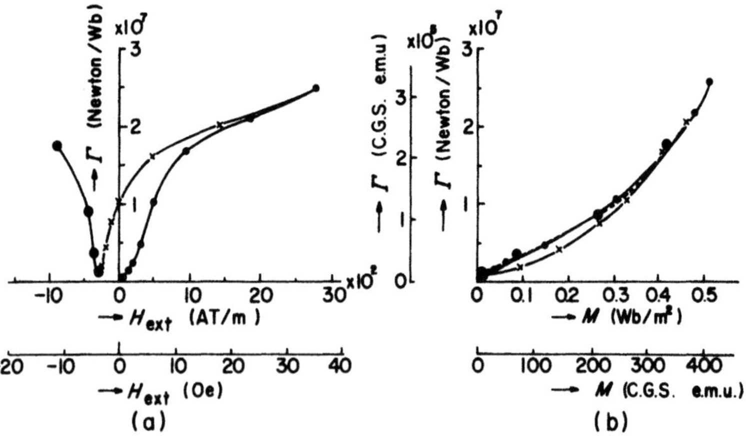

Fig. 19. Hysteresis in Γ characteristics; (a) Γ versus external bias field, (b) Γ versus bias magnetization.

6.1.5. Empirical Formula for the Magnetostriction of Nickel

Empirical formulas have been derived in 1942 by Kikuchi[56] for the magnetostriction phenomena of nickel. For the Joule effect, the strain is

$$\delta l/l = \alpha M^n \tag{50}$$

where $\alpha = 1.11 \times 10^{-4}$ mks (3.39×10^{-11} cgs) and $n = 2.246$, and the magnetostrictivity is

$$\Gamma = n\gamma M_{\text{bias}}^{(n-1)} \tag{51}$$

where $\gamma = 21.4 \times 10^6$ mks (65.8 cgs) which is in numerically good agreement with the relation

$$\gamma = \alpha E \tag{52}$$

when $E = 1.94 \times 10^{11}$ N/m². For material of higher internal stress, $n = 3$ has been suggested.[60]

6.1.6. Oxide Film as Insulation Coating of Nickel Sheet

In laminated nickel vibrators, oxide films, which have been formed on the surfaces of each sheet element during the annealing process in advance to lamination, are conveniently utilized as insulation coating. Kikuchi and Shimizu,[61] however, pointed out that the magnetostrictive characteristics may deteriorate considerably if the annealing temperature is inadequate, and recommended either of the following processes:

(1) For commercial pure nickel sheet of about 30% cold reduction, anneal at 700°C for 30 min in air.
(2) Insert hydrogen annealing at above 900°C before the film-formation process, which is the same as in (1).

6.2. Ni–Fe Alloys

In 1928, Schulze[62] published his detailed observation of static magnetostriction characteristics in this system of alloy throughout the entire range of atomic contents. The summarized data are shown in Fig. 20. The zero magnetostriction at 30% Ni content is due to the fact that the substance becomes nonmagnetic at this concentration. The zero around 81% Ni, however, is due to the so-called permalloy characteristics.

In 1937, Fukushima *et al.*[63] observed the dynamic characteristics of

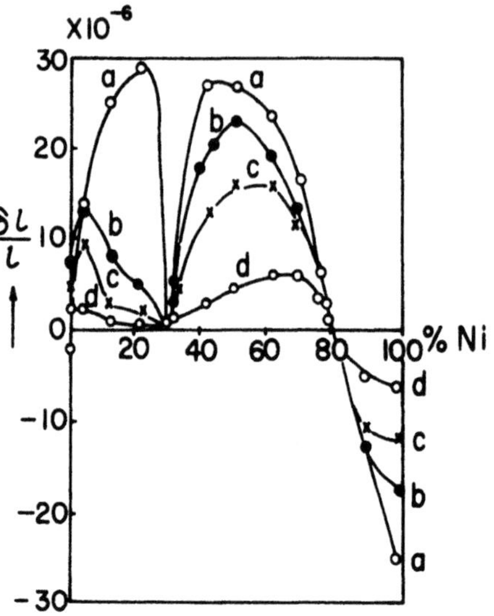

Fig. 20. Static magnetostriction of Ni–Fe alloys at various external magnetic fields: (a) 20,000 At/m (250 Oe), (b) 4000 At/m (50 Oe), (c) 2000 At/m (25 Oe), (d) 800 At/m (10 Oe) (after Schulze).

this system of alloy fabricated into 0.3-mm thin sheets. The data on the magnetostriction constant K are shown in Fig. 21(a).

Later in 1957, Davis, Helms, and Ferebee[64] observed the detailed magnetostriction properties of Ni–Fe alloys containing from 35% to 67.5% nickel, and similar results were obtained. The electromechanical coupling factor for alloys containing 40% to 52.5% nickel was greater than 0.31, and a maximum value of 0.37 was found for the 40% nickel alloy. Figure 21(b) shows the major constants observed at $H_{\text{bias}} = 640$ At/m (8 Oe) when the alloys are annealed at 1000°C in pure dry hydrogen. The performance of the alloys in this range of nickel content was equal to that of nickel for low-power transducer application, and particularly applicable when used in hydrophones. Pack annealing of the material at about 900°C was suggested.

6.3. Al–Fe Alloy

In 1928, Schulze[62] first observed the static magnetostriction of this system of alloy up to 10.52% of Al content. In 1941, Masumoto *et al.*[65,66] investigated a wider range of Al content varying from 0 to 16.65%, and

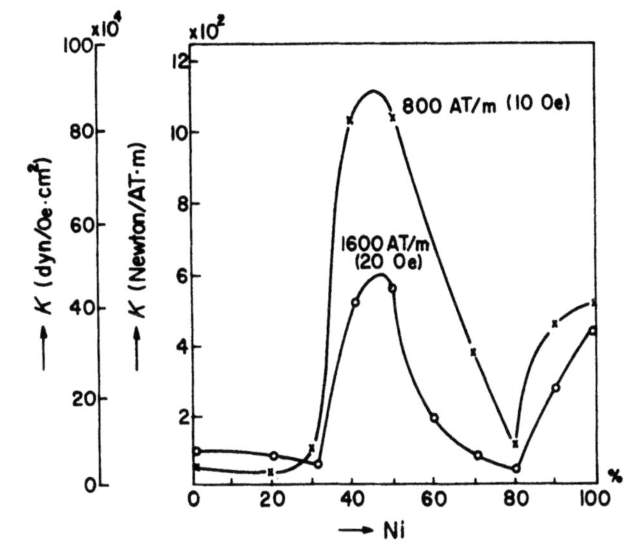

Fig. 21(a). Dynamic magnetostriction constant of Ni–Fe alloys at two bias magnetic fields, 800 At/m and 1600 At/m (10 Oe and 20 Oe) (after Fukushima *et al.*).

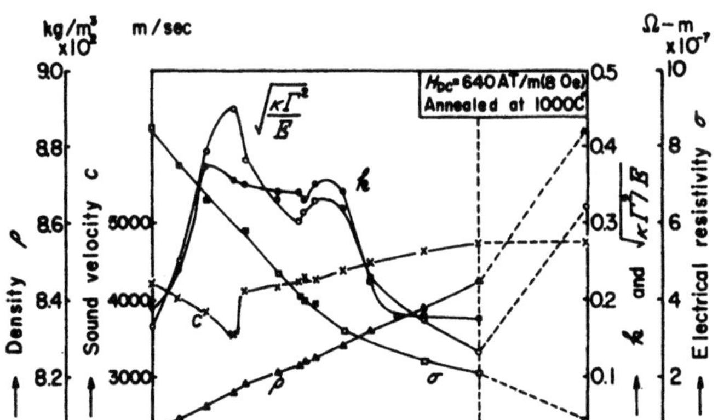

Fig. 21(b). Dynamic electromechanical coupling factor k of Ni–Fe alloys at 640 At/m (8 Oe), together with other major constants; c, sound velocity; ϱ, density; σ, electrical resistivity; and $\sqrt{\varkappa\Gamma^2/E}$, where $\varkappa$ is the reversible susceptibility, Γ the magnetostrictivity and E the Young's modulus in a relation $E = c^2\varrho$ (after Davis *et al.*).

found the so-called Alfer alloy, i.e., 11 to 13.5% aluminum–iron alloy, which has a static magnetostriction as high as that of nickel.

Following these findings, wide industrial investigations were rapidly carried out by a large organized group of researchers in Japan to develop the material in the form of thin sheets for transducer use. The Al–Fe alloy of 13 to 14% Al content, however, was so brittle that all the fabrication processes had to involve rolling at elevated temperatures. Figure 22 shows some basic characteristics of 0.4-mm sheet Alfer. $\varkappa\Gamma^2/E$ at optimum bias, which corresponds to the square of the usual electromechanical coupling factor, (*cf.* Section 2.3.5) reaches a maximum in the range of Al content coinciding with the range in which the static magnetostriction is the highest. In the figure, c is the propagation velocity of longitudinal elastic waves and ϱ is the density of the substance.

Later (1953–1955), this system of alloy was studied again in USA with a view towards easy fabrication,[67] taking the order–disorder transition temperature in to account. The rolling procedure consists of a small reduction at hot-rolling (1000°C) and a specific cold rolling at an elevated temperature of 575°C, which is slightly above the order–disorder temperature range. The material thus fabricated has been named Alfenol. Its use in transducers was discussed in detail by several authors.[68–70] One of the typical characteristics of Alfenol reported by Davis *et al.*[68] is shown in Fig. 23. The specimens are laminated rings of punched Alfenol sheet, 0.28 mm to 0.38 mm thick, mechanically resonating at about 30 kHz. Annealing took place at 1075°C in hydrogen. The 12- to 13-Alfenol is

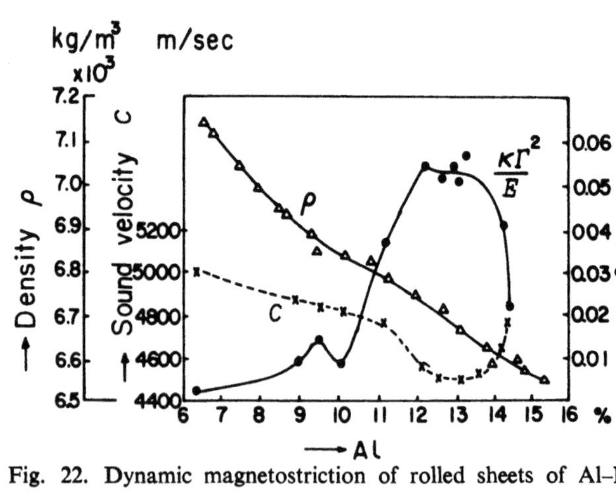

Fig. 22. Dynamic magnetostriction of rolled sheets of Al–Fe alloys at $H_{DC} = 800$ At/m (10 Oe); $\varkappa\Gamma^2/E$ is equal to k^2 (after Masumoto *et al.*).

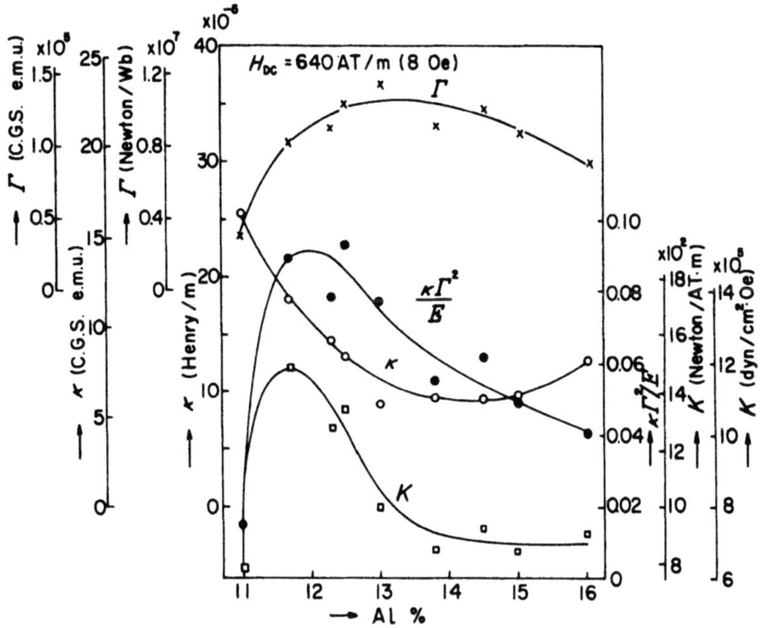

Fig. 23. $\varkappa$, Γ, K, and $\varkappa\Gamma^2/E$ of Alfenol as a function of aluminum content (after Davis *et al.*).

recommended for transducer application. As for K, however, 11.7-Alfenol has the maximum value. Pigott[69] has reported detailed investigation on the annealing temperature for 12.3- to 13.9-Alfenols and found that 12.3-Alfenol annealed at 650°C is the best for k^2 ($= \varkappa\Gamma^2/E$). Figure 24 shows k^2 of this material as a function of bias field. The maximum value is 0.12 and about twice as large as k^2_{max} for 13.4-Alfenol. As compared with that of soft annealed nickel, the value 0.12 is 25% higher. Among the specimens, 12.3-Alfenol was most sensitive to the annealing temperature and 13.4-Alfenol was almost insensitive to the temperature. The temperature range investigated was from 650 to 1062°C. The process of the annealing was as follows: The punched elements are heated in pure hydrogen at the specified temperature for 4 h and the cooling rate is regulated at 28°C/h down to 260°C. The additional comments in the report suggest that higher values of k^2 may exist for lower annealing temperatures and/or less aluminum content. Further studies have been made more recently in the USSR.[71,72] In one of the studies, the texture of this alloy was controlled by high rolling reduction and heat treatment to obtain high magnetostriction, up to 70×10^{-6} in the rolling direction.[73]

On the other hand, single crystals of Al–Fe alloy were investigated

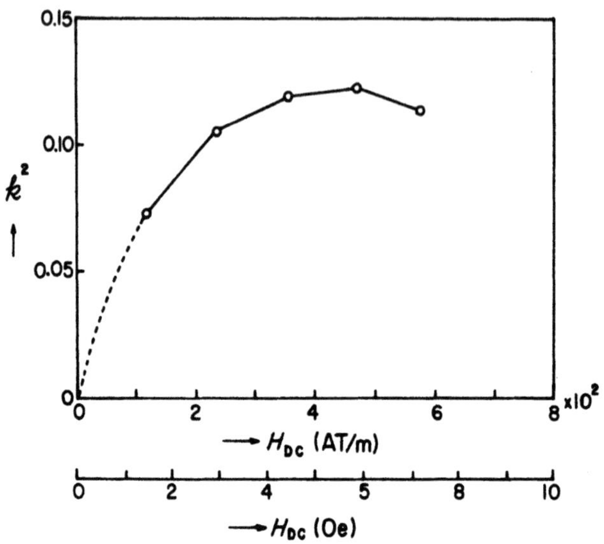

Fig. 24. Dependence of k^2 on polarizing magnetic field H_{DC} for the 12.3 Fe–Al alloy annealed at 650°C (after Piggot).

for their magnetostriction in various ways. In 1957, Hall[74] measured the spontaneous saturation magnetostriction and the forced magnetostriction of slowly cooled single-crystal Al–Fe alloys in the composition range from 6 to 30 at% aluminum. In 1958, Gengnagel[75] observed the linear magnetostriction of single crystals of this system in relation to their heat treatment to show its clear change between ordered and disordered lattices.

6.4. Other Metals

In 1939 and 1940, Nukiyama *et al.* investigated Fe–Si–Al alloys in the composition range up to 4% Si and 4% Al.[76] The magnetostriction constant K of the alloys in this range was as large as that of nickel because of their high permeability, but the electromechanical coupling factor was too small for use in general ultrasonic transducers because of their low values of Γ. In 1944, Kikuchi *et al.*[32] investigated another alloy in the system of Fe–Cr–Al with some addition of W and Si, which was characterized by comparatively large coupling factors ($\varkappa\Gamma^2 = 2.6 \times 10^9$ N/m²) when used at its magnetic remanence as the bias. The optimum composition for industrial use was specified at 20% Cr and 2.6% Al, and the alloy was named Cromal.

Dynamic magnetostriction characteristics of grain-oriented silicon steel

(Hipersil type) were studied in 1943 by Kikuchi.[77] The value of $\varkappa\Gamma^2$ at optimum bias was about 0.3×10^9 N/m^2 in the direction parallel to the rolled direction and 0.05×10^9 N/m^2 in the perpendicular direction. Clear anisotropy was also found for the propagation velocity of longitudinal elastic waves, namely, 5200 m/s in the parallel direction and 4500 m/s in the perpendicular direction.

Sussman and Ehrlich[26] reported in 1950 the evaluation of the magnetostrictive properties of an alloy of 0.4–0.55% Cr, 34.5–35.5% Co with the balance Fe, which had been named Hiperco. Figure 25 shows the electromechanical coupling factor as a function of biasing flux density. The coupling factor reaches 0.17 at around 1.5 Wb/m^2 (15,000 gauss). The effective use of magnetic remanence for the bias is the main advantage of this alloy, because the characteristics at remanence ($\varkappa\Gamma^2 = 4 \times 10^9$ N/m^2 at remanence) are as good as Permendur, but inferior to nickel and Permendur when used as a high-power ultrasonic transmitter in liquid.

Nickel–cobalt base alloys have been studied in various ways by many researchers for their magnetostrictive properties. The dynamic magneto-

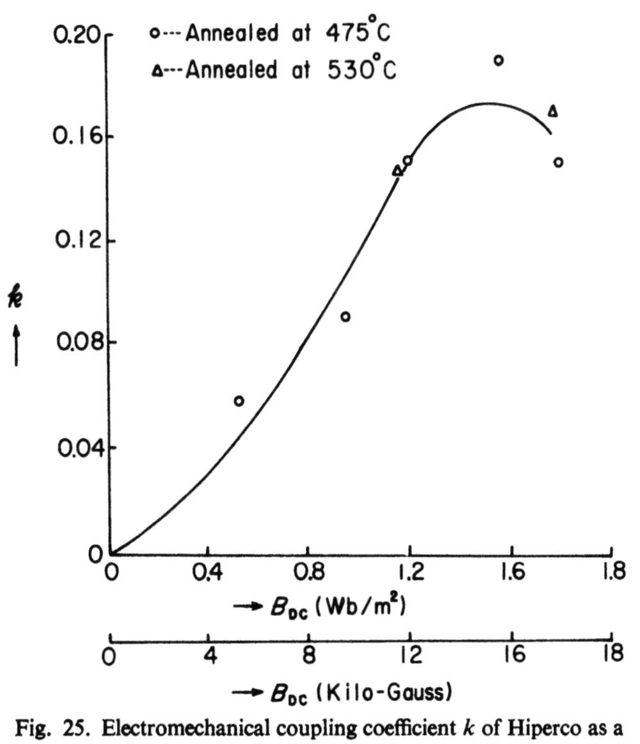

Fig. 25. Electromechanical coupling coefficient k of Hiperco as a function of flux density B_{DC} (after Sussman *et al.*).

striction of nickel–cobalt alloys were discussed by Clark[78] in the range from 0 to 23% cobalt in nickel. Alloys containing 4.4 and 18.4% cobalt were recommended for use both as transmitting and receiving transducers because of their superiority to the other magnetostrictive materials including nickel. Their higher Curie point, the ease of rolling, and high corrosion resistance were pointed out as additional advantages. An electromechanical coupling factor of 0.5 was claimed with 4.5% cobalt–nickel alloy.[79] For the use in the higher frequency range, a ternary alloy containing 1.4% cobalt, 2.3% chromium, with the balance nickel[80] has also been suggested since this alloy has a higher electrical resistivity ($0.3 \times 10^{-6} \Omega \cdot m$) though the coupling factor decreases from 0.5 to 0.37. The temperature dependence of the magnetostrictive properties of the ternary alloys was determined over the range from −60 to +60°C. An almost temperature-independent mechanical resonance was observed in an alloy of 2.3% Cr, 1.4% Co, with the balance Ni.[81] A 65% cobalt–iron alloy was investigated[82] for its magnetic and magnetostrictive characteristics, showing certain advantages and disadvantages compared to Ni, Alfer, and Permendur.

Whymark[52] observed the effect of compressive stress on the dynamic magnetostriction characteristics of a 2V-permendur sheet, which was 0.125-mm thick. The specimen was annealed in hydrogen at about 575°C

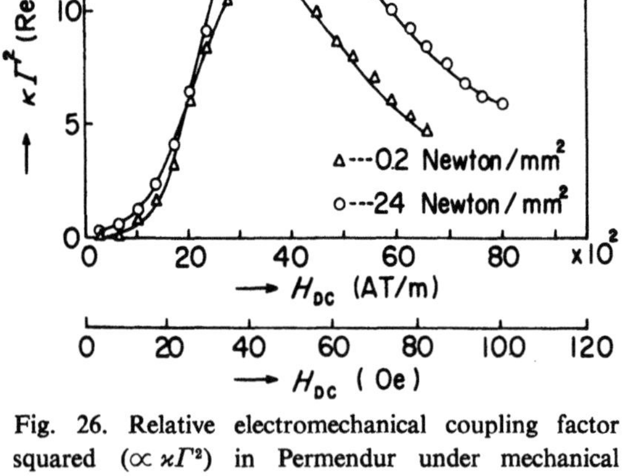

Fig. 26. Relative electromechanical coupling factor squared ($\propto \kappa\Gamma^2$) in Permendur under mechanical compressive stress (after Whymark).

for 80 min. The effect on its electromechanical coupling is shown in Fig. 26. When the compressive stress is raised up to 24 N/mm² (240 kg/cm²), the maximum of the coupling coefficient increases considerably, in contrast to the effect in nickel (see Fig. 18). Whymark commented that the effect was not an unexpected result because nickel shrinks when the magnetic field is increased, while permendur expands (*cf.* section 2.6).

6.5. Cobalt Rondel

The magnetostriction of low-density cobalt rondel was observed in 1958 by Strauss.[83] Figure 27 shows its high magnetostrictive strain (-139×10^{-6}) at the magnetic induction of 0.7 Wb/m² (7000 gauss), whereas metallic cobalt of normal density develops a strain of only -28×10^{-6} at about 1.8 Wb/m² (18,000 gauss). The rondels are porous and have less than 50% of the density of metallic cobalt. The rondels from which the specimens were cut were small cylinders commercially produced by reducing a sintered cobalt hydrate or black oxide.

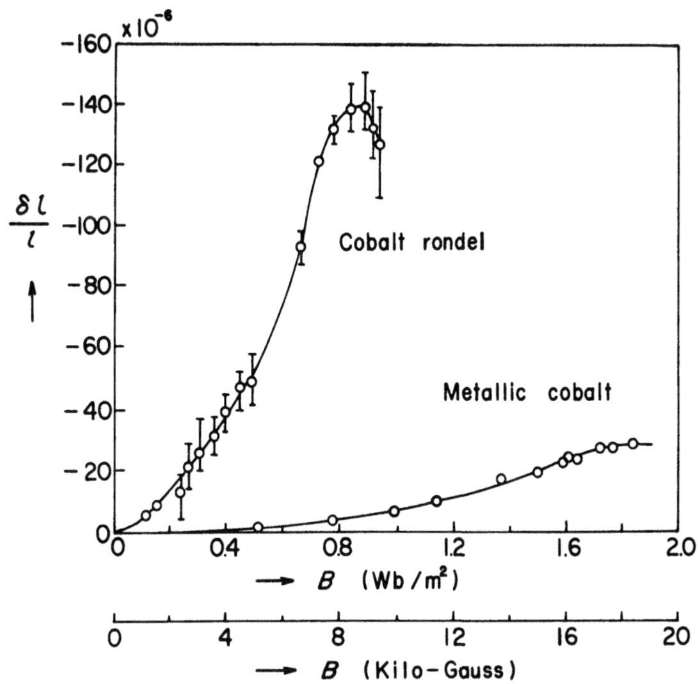

Fig. 27. Static magnetostriction of cobalt rondel and metallic cobalt as a function of magnetic induction *B* (after Strauss).

6.6. Ferrites

Magnetostriction phenomena of ferrite substances were first reported by Takei and Nakamura in 1943,[84] and also by Van der Burgt in 1953 with regard to Ni–Zn ferrites.[85] Thiede[86] described in 1954 a prototype ultrasonic vibrator made of ferrite. Kikuchi *et al.*[8,9] in 1955 published data on the Ni–Cu system together with some types of ferrite vibrators biased by ferrite permanent magnets (see Fig. 2). They further studied the magnetostrictive properties of Ni–Zn ferrites,[87] Ni–Cu ferrites,[88] Ni–Cu–Co ferrites,[89] Ni–Cu–Zn–Co ferrites,[90] and other ternary, quaternary, and quinary ferrites of nickel–ferrite base by taking the ferrites of Cu, Co, Zn, Cd, V, Cr, Li, and W in various combinations as the additional ingredients. The investigation also covered the effect of various additional ingredients such as MgO, SiO_2, MnO, Na_2O, K_2O, Al_2O_3, TiO_2, As_2O_3, CaO, BaO, SrO, PbO, Sb_2O_3, etc., either intentionally added in small amounts or present as impurities.

Independent of the above work, extensive research on magnetostriction properties of ferrites was carried out by Van der Burgt[91] on Ferrox-cube (Philips ferrites) with examples of ultrasonic applications.

As will be shown in later sections, magnetostrictive ferrites must have a high mechanical strength for most applications. This requirement is often more important than such properties as electromechanical coupling factor and temperature coefficient of the elastic modulus. The proper choice of the chemical composition in industrial magnetostrictive ferrite, requires a compromise. After extensive experiments on an industrial scale Kikuchi *et al.* have chosen certain chemical compositions of Ni–Cu–Co ferrite. This material is now commercially available in Japan and widely used in various applications.

The chemical composition of the current industrial ferrites is given by the formula

$$\{(NiO)_x \cdot (CuO)_{(1-x)}\}_{(1-y)} \cdot (CoO)_y \cdot (Fe_2O_3)_1$$

where $x = 0.85$–0.9 and $y = 0.004$–0.015. In this range, the ferrite having less copper and more cobalt is called Vibrox-1. It has a larger electromechanical coupling factor (k_{max}) at optimum magnetic bias. The ferrite having more copper and less cobalt is called Vibrox-2 and has a mechanical resonant frequency which is less temperature dependent (determined by the elastic modulus) and at the same time has higher mechanical strength. Figure 28 shows the temperature coefficient of the resonant frequency as a function of the chemical ingredients. The characteristics shown in this

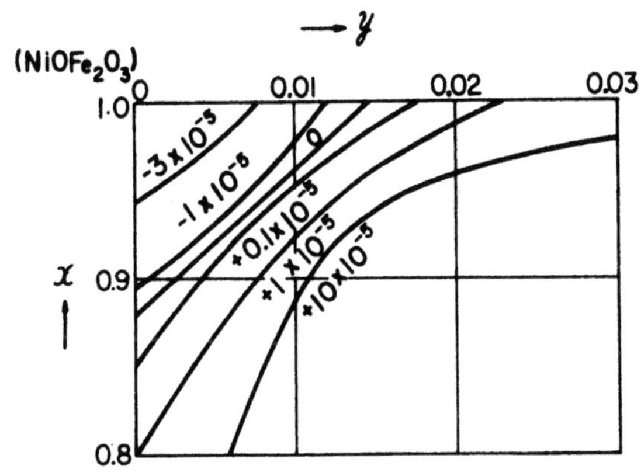

Fig. 28. Temperature coefficient of resonance frequency of Ni–Cu–Co ferrite magnetostrictive vibrator (20° to 100°C).

figure may change slightly depending on the production process. Figure 29 shows the typical characteristics of k_{max} in relation to the two ingredients. The mechanical strength, however, is greatly affected by various factors in the manufacturing process such as preliminary firing, ball-milling, molding, drying, final firing, etc. These factors also affect the other two major requirements to a certain extent.

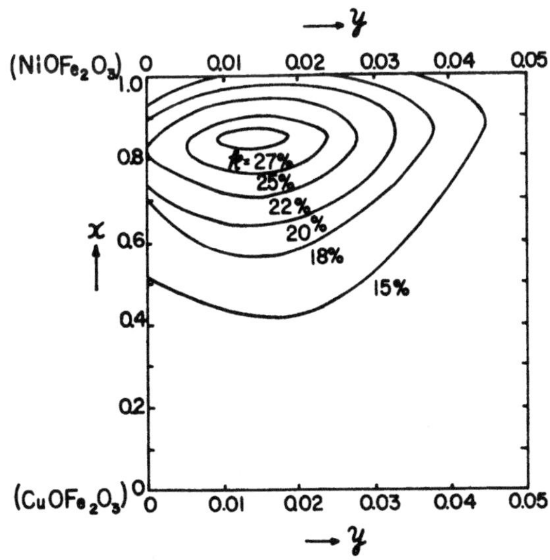

Fig. 29. Electromechanical coupling factor k of Ni–Cu–Co ferrite at optimum bias.

The mechanical strength largely depends upon the accuracy of the manufacturing process. Recent examinations[92] show that a value of 30×10^6 N/m^2 (300 kg/cm^2) is usually attained, and that if a certain process of selection is applied to the production line, vibrators which are guaranteed to 40×10^6 N/m^2 (400 kg/cm^2) are industrially available in production quantity. The guaranteed stress level is set lot by lot at half the ultimate strength dynamically observed in each production lot by breaking tests of a certain number of samples of the actual vibrators to be sold.

Many other reports on magnetostriction in ferrites have been published in the last ten years. Ferebee and Davis[93] reported in 1958 on Ni–Co–ferrous ferrites having electromechanical coupling factors of 0.28–0.37 and stated that the maximum stress increased with ferrous ion content. Sokolov and Shur[94] discussed the magnetostrictive sensitivity in Ni–Zn ferrites, using a relation, $e_{\max} \propto \mu_0 \lambda_s / I_s$, where $e_{\max}$ is the output voltage for a specified acoustic pressure, μ_0 the initial permeability, λ_s the saturation magnetostriction, and I_s is the saturation magnetization. Golyamina[95] discussed such ferrite materials as Ni, Ni–Zn and Ni–Co ferrites for acoustic transducers. Smith and Mattiat[96] discussed the controlled temperature dependence of Ni ferrite when Ni is substituted by Cu and Co and its effect on the reproducibility of properties. Van der Burgt and Stuijts[97] discussed the magnetostrictive characteristics of low-porosity ferrites for high-intensity use. Koczkowski[98] studied the hysteresis in the magnetomechanical parameters of Ni–Co–Mn ferrites. Golyamina[99] discussed Ni–Zn ferrites with various compositions for the use as ultrasonic transmitters. Ganeva and Golyamina[100] studied the magnetostrictive properties of Ni, Ni–Zn, and Ni–Co ferrites at high temperature up to the Curie point, where magnetostriction disappeared. $NiO \cdot Fe_2O_3$ showed the highest Curie point (590°C), and $NiO_{0.5} \cdot ZnO_{0.5} \cdot Fe_2O_3$ showed the lowest (260°C). Kikuchi *et al.*[101] observed the high-temperature characteristics of Vibrox-2 (Ni–Cu–Co ferrite) transducers and indicated that no appreciable change of dynamic characteristics were observed up to 300°C, and that the characteristics were reversible up to 400°C, when the transducer is biased with a barium ferrite permanent magnet, and reversible up to over 500°C when the transducer is biased by direct current. Figure 30 is an example of the latter case, showing the temperature dependence of the characteristics and material constants derived from the observation on an actual ferrite vibrator of rectangular type; f_0 is the resonant frequency, A the force factor (A-type), Q the mechanical quality factor of resonance, k' and $\varkappa$ are the electromechanical coupling factor and dynamic magnetic susceptibility, respectively, which are apparent on the material of the vibrator, obtained through an approximate

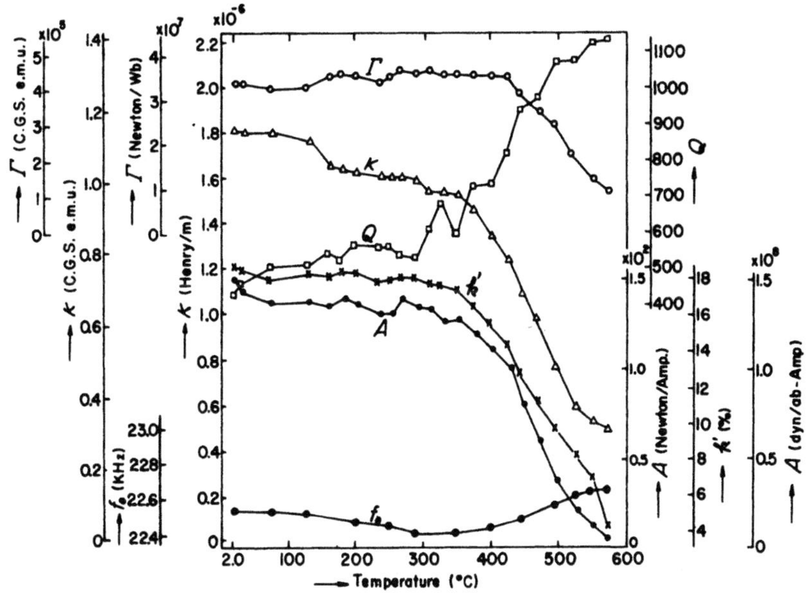

Fig. 30. Temperature dependence of the characteristics and material constants in Vibrox-2 (Ni–Cu–Co ferrite).

analysis, and Γ is the magnetostrictivity. The biasing field was kept at a constant value, which is an optimum at room temperature. The B-type force factor(31,22) of an actual vibrator made of Vibrox-2 (Ni–Cu–Co ferrite) was first observed in 1958 by Kikuchi *et al.*(102) up to a stress amplitude of 14×10^6 N/m^2 (140 kg/cm^2). The force factor decreases for amplitudes above 5.6×10^6 N/m^2 (56 kg/cm^2) (see 7.2.4). The decrease, however, can be compensated even at the stress amplitudes of 21.4×10^6 N/m^2 (214 kg/cm^2)(103) when larger bias magnetization is applied. In 1965, Golyamina and Chulkova(104) also suggested the application of higher bias while describing the behavior of large signals in various ferrites (see also 7.2).

6.7. Theoretical Models for the Characteristics of Magnetostriction in Polycrystalline Metals*

Kikuchi(105) proposed in 1943 a synthetical method for predicting theoretical characteristics of magnetostriction in polycrystalline metals by introducing such a concept as contribution probability p_i of the in-

* The following sections (6.7 to 7.1) discuss the theoretical aspects of magnetostriction phenomena. For a discussion of the experimental aspects see Section 7.2.

dividual microcrystals which compose the polycrystalline metals. The shape distribution of the microcrystals was to be taken into account in p_i together with the probability of the crystal orientation. The subscript i means the crystallographical direction of the microcrystal in relation to the lengthwise direction of the polycrystalline body, in which magnetostriction is under consideration.

6.7.1. Static Magnetostriction

The static magnetostriction characteristics $\xi'(M)$ can be predicted by Eq. (53), where the $\xi_i'(M)$ are those for the microcrystals, i.e.,

$$\xi'(M) = \sum_{i=1}^{n} p_i \xi_i'(M) \tag{53}$$

If such theoretically obtained characteristics as Heisenberg's formulas[106] or Gans and Harlem's formulas[107] are applied to $\xi_i'(M)$ the curves in Fig. 31 are obtained showing an example for polycrystalline pure iron of uniform grade.[105] As there are half empirical constants λ_{100} and λ_{111} in Heisenberg's formulas, Honda and Mashiyama's experimental data[108] are employed in the calculation of Curve A, and Webster's data[109] in Curve B. The dots in the figure are certain experimental data. With regard to pure nickel, similar results were obtained.

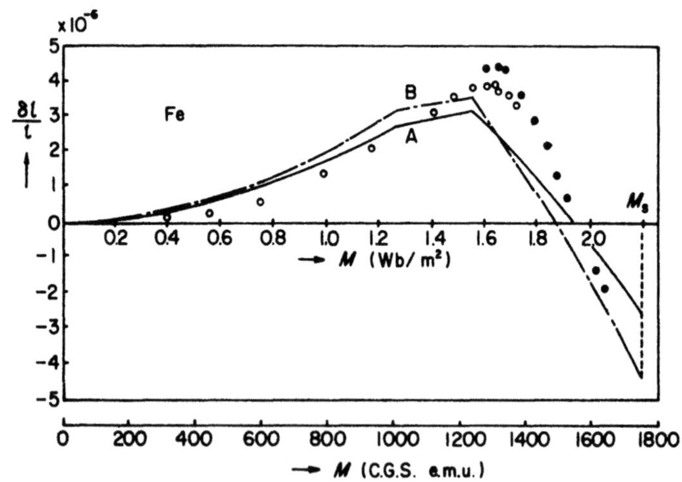

Fig. 31. Theoretically synthetized characteristics of static magnetostriction in polycrystalline pure iron of uniform grade (plots being the experimental data for comparison).

Later, Callen and Goldberg[110] discussed the magnetostriction of polycrystalline aggregates by reinterpreting previous data and mathematical treatments on crystals. Their calculation, however, applies only to saturation magnetostriction.

6.7.2. Dynamic Magnetostriction

Kikuchi extended his *a priori* method also to the dynamic characteristics,[105] enabling us to synthetize the modulus of elasticity E and the magnetostrictivity Γ of polycrystalline metals. The characteristics Γ of the polycrystal as the function of bias magnetization M_{bias} is expressed by the following formula:

$$\Gamma(M_{\text{bias}}) = \left[\sum_{i=1}^{n} p_i \frac{\partial \xi_1'(M)}{\partial M} \Big/ \sum_{i=1}^{n} p_i \frac{1}{E_i} \right]_{M=M_{\text{bias}}} \tag{54}$$

The synthetized results for pure iron and pure nickel are in a good agreement with experimental data obtained dynamically (by magnetostrictive vibration) with sheet iron and nickel respectively. In the synthesis, for E_i of single crystal pure iron, Kimura and Ohno's data[111] are employed, and for that of pure nickel, Honda and Shirakawa's[112] and Yamamoto's[113] are employed.

6.7.3. Anisotropic Grain Orientation

These methods of the synthesis have been extended towards problems on grain-oriented sheet materials for magnetostriction use,[105] and discussed the anisotropic magnetostriction characteristics of Hipersil-like materials as the function of intensity of magnetization.

Later, in 1963, Langham[114] discussed the static magnetostriction of polycrystalline silicon–iron sheet, having Goss (1935) texture, with a model based on the orientation distribution of the microcrystals and domain theory of magnetization. Langham claimed good numerical agreement with experimental data.

7. CONSIDERATION OF LARGE SIGNAL OPERATION

7.1. Theoretical Approach

In 1946, Kikuchi[115] proposed a theoretical method for estimating the limit of mechanical output power available from magnetostrictive vibrators. The theory is based on the fact that any material can be used for magneto-

Table III. Major Magnetostrictive Materials (MKS units)

	Chemical composition %	Static magnetostriction saturation $(\delta l/l)_s$, $\times 10^{-6}$	DC bias H_{DC}, At/m	Magnetostrictive activity $\varkappa(0)\Gamma^2$ 10^9 N/m²	Electromechanical coupling factor k %	Dynamic relative permeability,[a] $\mu_s(0)$ at a bias H_{DC} (At/m)	Specific electrical resistance σ $10^{-6}\ \Omega \cdot m$	Sound velocity of a long thin bar c 10^3 m/s	Density ϱ 10^3 kg/m³
Metals									
Pure Nickel (soft, 900°C H_2 annealed)	Ni > 98	−40	800∼1600 (10∼20 Oe)	5∼13	16∼25	60 (1200)	0.07	4.75∼4.90	8.7
			at remanence	2.9	12	50	0.07	4.75	8.7
Pure Nickel (semi-hard, 600°C H_2 annealed)	Ni > 98	−35	at remanence	4	14	19	0.098	4.75	8.9
Alfer[66] (AF Alloy)	Al 13, Fe bal	40	500∼800 (6∼10 Oe)	5	17∼20	190 (800)	0.91	4.30∼5.20	6.7
Hiperco[26]	Cr 0.4∼0.5, Co 34.5∼35.5, Fe bal	40	B_{DC}=1.5 Wb/m² (15,000 gauss)	6	17	75	0.23	5.08	8.1
			at remanence	4	14	123			
Hipernik	Ni 50, Fe bal	25	560∼1200 (7∼15 Oe)	6	20.5	190 (800)	0.46	4.10∼4.20	8.3
40% Permalloy	Ni 40, Fe bal	25	600∼1200 (7.5∼15 Oe)	4	17	230 (600)	0.75	4.00∼4.10	8.2

Fe–Co Alloy	Co 70, Fe bal	75	1450 (18)	1.5	—	90 (1450)	—	4.80	—
Cromal[32]	Cr 20, Al 2.6, Fe bal	28	at remanence	2~2.6	10~11.5	≃440	1.20	5.00	8
Ni–Cr Alloy	Cr 4, Ni bal	−8	400 (5)	8	—	100 (400)	0.40	4.85	—
Al–Si–Fe Alloy	Al 4, Si 2, Fe bal	11	320 (4)	0.5	5.5	200 (320)	0.80	4.90	7
Pure Iron (electrolytic)	—	−9	1600~2400 (20~30)	—	—	50 (2400)	0.10	5.00	7.8
Ceramics							$\Omega \cdot m$		
Vibrox-1	Ni–Cu–Co Ferrite	−30	1200~1600 (15~20)	8	22	12 (1600)	>4	5.60	5.1
Vibrox-2	Ni–Cu–Co Ferrite	—	1200~1600 (15~20)	11	27	22 (1600)	>4	5.40	5.1
Ferrox Cube[125]									
7A1	Ni–Cu–Co Ferrite	−28	—	—	25–30	—	1–10	—	—
7A2	Ni–Cu–Co Ferrite	−28	—	—	22–25	—	1–10	—	—
7B	Ni–Co Ferrite	−27	—	—	19–22	—	$>10^2$	—	—
4E	Ni Ferrite	−27	—	—	18–21	—	$>10^2$	—	—

[a] $\mu_s(0) \equiv (\mu_s)_{\omega=0}$ in a general relation of $\mu = \mu_s\mu_0$, where μ_0 is the permeability of vacuum. $\varkappa$ can be obtained by a relation $\varkappa = (\mu_s - 1)\mu_0$. Value for $\mu_s(0)$ is numerical and H_{DC} is given in ( ).

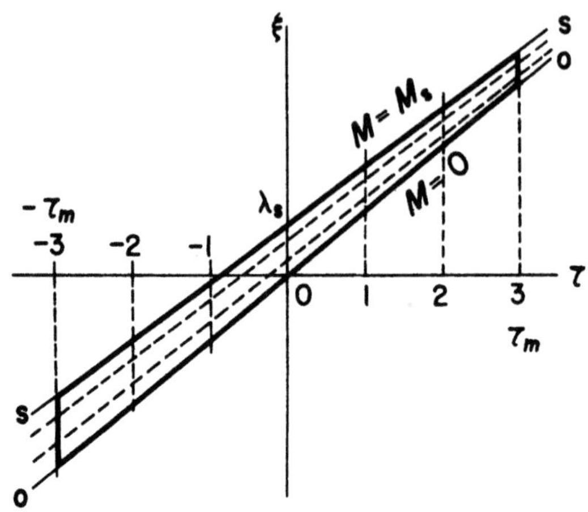

Fig. 32. Stress–strain plane expressing the physical state of a substance for large amplitudes of stress and magnetization.

strictive vibrators only within the following two limits: (1) The magnetic saturation, which limits the available maximum driving force, and (2) the mechanical fatigue limit of dynamic stress, τ_m, in the material. The quadrilateral* in τ versus ξ shown in Fig. 32 is the region in which the material with isotropic magnetostriction (e.g., nickel) may be operated as a vibrator. If the physical state of the substance is varied so that it draws a locus which just coincides with this quadrilateral, the available mechanical output is the maximum. The area A_m of the quadrilateral is equal to the mechanical work done by the unit volume of the substance in one cycle of the physical state. For sinusoidal phenomena the fundamental component of the output work is approximately equal to the area A_{m_1} of an ellipse inscribed to the quadrilateral. The area A_{m_1} is

$$A_{m_1} = \pi A_m/4 = \pi \lambda_s \tau_m/2 \tag{55}$$

In order to obtain maximum output, the mechanical load must be chosen so that the alternating mechanical stress in the substance will reach the value τ_m when the alternating magnetization changes between zero and M_s. This condition is called "activity matching." Generally, activity matching is not obtained, and there are two cases: (i) The mechanical load is so light that the alternating stress reaches the allowable (fatigueless)

* The $\xi - \tau$ characteristics are not exactly straight lines, especially at large stresses. According to certain theoretical considerations[41] quadrilaterals can be approximated.

limit before the alternating magnetization reaches saturation. This case is called "under matching." (ii) The mechanical load is so heavy that the alternating stress is less than the allowable limit even when the alternating magnetization reaches saturation. This is the case of "over matching." In either mode of operation, the area of the working ellipse is obviously smaller than A_{m_1}. Therefore, the ultimate output available from a magnetostrictive vibrator is usually less than the potential maximum output of the material.

In 1957, Kikuchi *et al.*[116] improved the theory so that it is applicable to various actual cases. In the following, the outline will be described; for the details and curves for practical designing calculation, see Ref. (7) or Ref. (117). Nearly all actual vibrators, such as shown in Fig. 1(a), working in a liquid, are under-matched. In this case, the ultimate sound intensity P/S (output power P from a rectangular vibrator, divided by the radiating area S) is

$$P/S = \tfrac{1}{2}\zeta \varrho_a c_a c_i^2 \xi_m^2 \mathscr{P}_u(\omega) \tag{56}$$

In this formula, $\mathscr{P}_u(\omega)$ is a form factor determined by the physical configuration of the vibrator (usually, $\mathscr{P}_u(\omega) \approx \frac{1}{2}$), ξ_m is the allowable maximum strain in the material, c_i is the propagation velocity of longitudinal elastic waves in the material, $\varrho_a c_a$ is the characteristic acoustic resistance of the medium, and ζ is a cavitation coefficient representing the decrease of the radiation resistance due to cavitation at the radiation surface of the vibrator.[118,119,120] The value of ζ ranges from unity, in a noncavitating liquid, to 0.15, when cavitation is very pronounced. For example, for annealed nickel[121] ($\xi_m = 200 \times 10^{-6}$) and for the usual rectangular vibrator of $q = 0.5$ (see Fig. 7) the value of P/S is 5 W/cm^2 at intense cavitation in water; for Vibrox-2 (Ni–Cu–Co ferrite)* ($\xi_m = 250 \times 10^{-6}$) P/S becomes 10 W/cm^2.

For optimum acoustic loading where the vibrator operates in an activity-matching condition the ultimate output intensity is

$$P_m/S = \tfrac{1}{2}\eta_m E \lambda_s c_i \xi_m \mathscr{P}_m(\omega) \tag{57}$$

where $\mathscr{P}_m(\omega)$ is another form factor, having the value of around 0.3 for the usual rectangular vibrators, and η_m is the mechanoacoustic efficiency.

* This value is based on the assumption that $\tau_m = 40 \times 10^6$ N/m^2 (400 kg/cm^2), as stated in the foregoing section on ferrites, and that the modulus of elasticity, E, is 1.6×10^{11} N/m^2 (1.6×10^6 kg/cm^2).

For vibrators made of nickel, P_m/S has been estimated to be 77.8 W/cm^2 when η_m is assumed to be 0.8. The acoustic load for this mode of activity matching must be about 2.3 times larger than that of the noncavitating water.

Further theoretical studies have been made on the wave forms of large amplitude magnetization and magnetostrictive driving force, magnetic hysteresis loss in the state of superimposed large amplitude mechanical stress of the same frequency,[45,122] electroacoustic transduction efficiencies, etc., including the cases of a nonbias excitation (so-called half-frequency excitation).

7.2. Experimental Approach

7.2.1. Calorimeter Method

Various experimental methods have been proposed for the consideration and treatment of large signal operation. Whymark[123,124] used a calorimetric method for observing intense sound output of magnetostrictive vibrators together with a graphical method for obtaining dynamic magnetostrictive stress. The matching adjustment of acoustic load was suggested to attain saturation output. The measured maximum in nickel was reported as being 167 W/cm^2. The outline of the proposed mechanical system is as follows: The magnetostrictive specimen to be observed in its longitudinal fundamental mode is freely supported at one end of a resonant stub as shown in Fig. 33. The length of the stub is one wavelength and each of its two velocity nodes are supported by a thin diaphragm. At the other end of the stub, an impedance transforming solid horn is coupled to step up the vibrational velocity, which supplies acoustic power to a calorimeter apparatus. The power absorbing part of the calorimeter is a cast-iron stub a half-wavelength long. A pair of suitable couplers are used for the water inlet and outlet of the calorimeter. The vibrational velocity distribution of the whole system is shown in the lower part of Fig. 33.

The specimen is magnetostrictively driven by an exciting coil which is wound around the specimen with a suitable magnetic yoke. The magnetic state of the specimen is observed by a search coil and various instruments. The vibrational velocity of the specimen is directly observed by an capacitance-type ultramicrometer. Figure 34 shows the experimental results for nickel obtained by Whymark. The dotted curves represent the magnetostrictive driving force calculated from the static curve as a function of the alternating magnetization which is assumed to be sinusoidal, while the

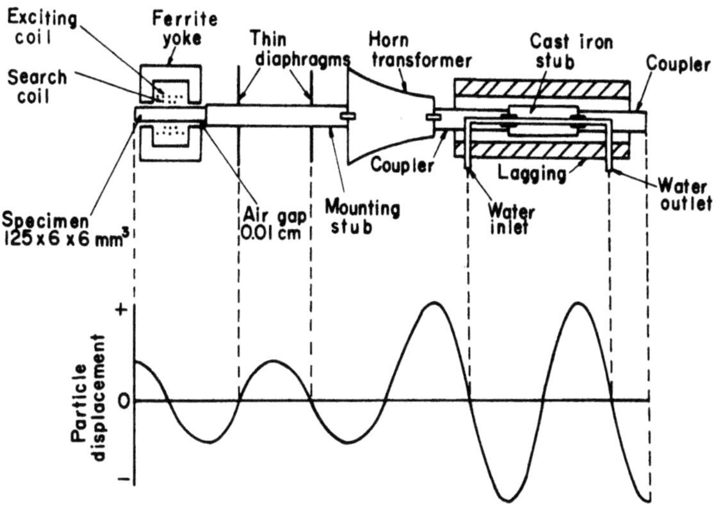

Fig. 33. Mechanical system of Whymark's calorimetric method for measuring intense sound output of magnetostrictive vibrators. The lower diagram shows the distribution of particle displacement along the composite mechanical structure.

solid curves are obtained by the experimental determination with the specimen, the parameters on each curve being the polarizing field. The quiescent polarization, which is given by a polarizing field, is not necessarily equal to the biasing magnetic polarization. This fact is due to the hysteresis and nonlinearity of the minor loop.

7.2.2. Vibrometer Method

Kikuchi *et al.*[102] described two methods for measuring the force factor of a magnetostrictive vibrator while it was operating at large amplitudes. One is a vibrometer method and the other a differential method. The details for the latter will be described later. The vibrometer method uses cascade bonding of a velocity-measuring vibrator to the vibrator under observation.

In Fig. 35a, S_1 is the magnetostrictive transmitter whose operation is to be observed. On the top surface of S_1 another vibrator S_2 of the same resonant frequency is cemented in such a manner that the windings of the two vibrators have no electromagnetic coupling. Then, the vibrational velocity of the transmitter can be measured by vibrator S_2. The bottom surface of the vibrator S_1 is the radiating surface, which is immersed into an acoustic medium. This combination is called a vibrometer. Figure

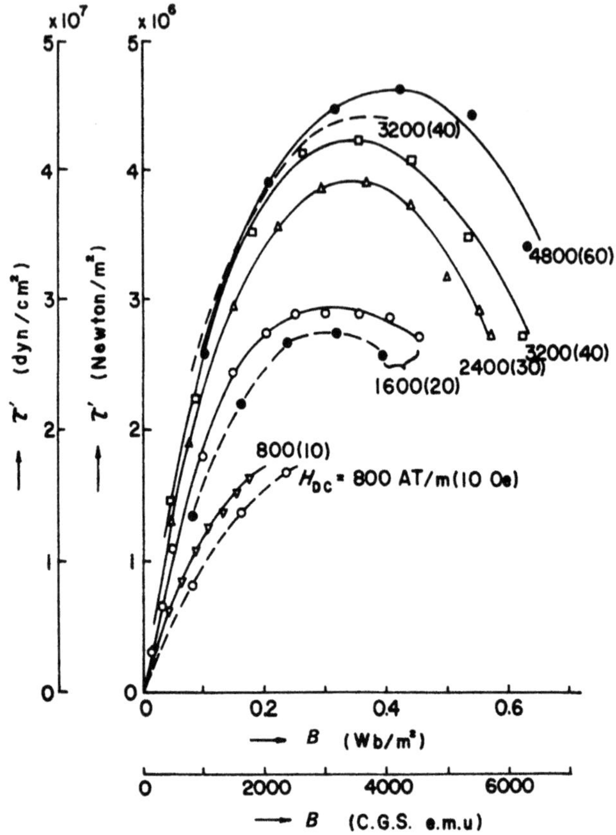

Fig. 34. Dynamic magnetostrictive driving force for annealed nickel as a function of alternating magnetization B (peak value). The dotted curves, calculated from static and quasi-static measurements; the solid curves, resonance determination (after Whymark).

35b shows an other vibrometer configuration. The dotted line in the figure indicates an imaginary plane separating S_1 and S_2. No magnetic coupling exists between the two because of the flux cancelling effect. Figure 35c shows a simplified electrical circuit diagram of the vibrometer.

The fundamental electroacoustic equations of this vibrometer system can be written as follows:

$$I_1 = Y_{d1}V_1 - A_1 v \tag{58a}$$

$$I_2 = Y_{d2}V_2 - A_2 v \tag{58b}$$

$$F = -A_1 V_1 - A_2 V_2 + (z_a + z_0)v \tag{58c}$$

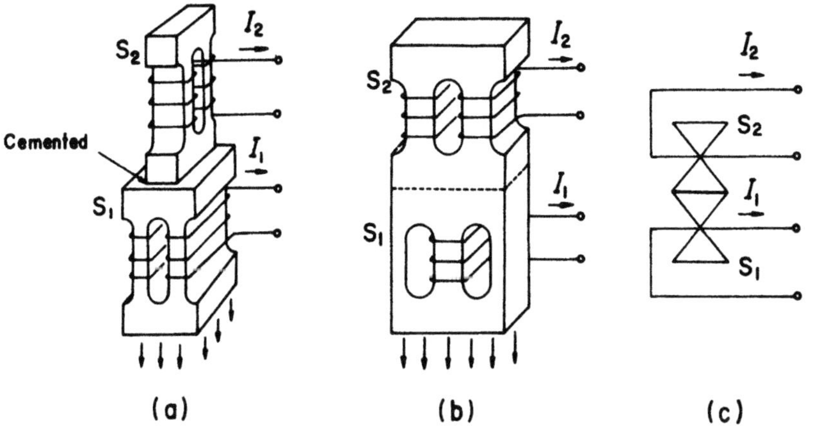

Fig. 35. "Vibrometers"; (a) and (b), magnetostrictive vibrometers; (c), its simplified expression in an electrical circuit diagram.

where the subscripts "1" and "2" refer to the winding of the vibrators S_1 and S_2 respectively; I_1 and I_2 are the electric currents in the windings, V_1 and V_2 are the voltages, Y_{d_1} and Y_{d_2} are the clamped admittances, A_1 and A_2 are the force factors, F is the external mechanical force at the acoustic radiating surface, v is the vibrational velocity of the radiating surface, z_a is the acoustic radiation impedance, and z_0 is the internal mechanical impedance of the vibrometer. The principle of the vibrometer method is shown in Fig. 36. A high-frequency current drives the S_1-vibrator at its mechanical resonance to generate intense ultrasonic vibrations; the winding of the S_2-vibrator is short-circuited so that V_2 is equal to zero. Then the short-circuited current I_m of the S_2-vibrator is

$$I_m = -A_2 v \tag{59}$$

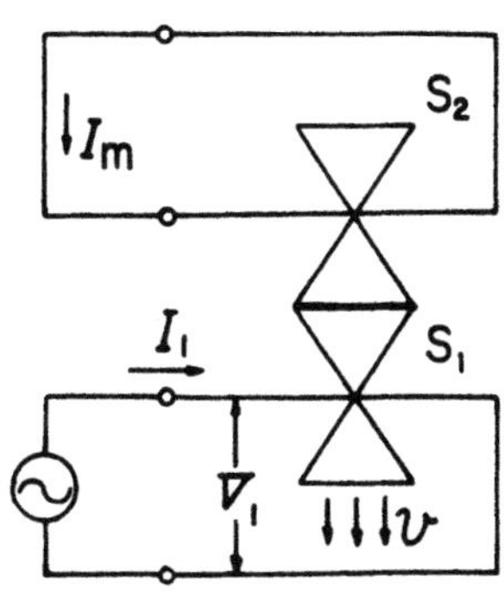

Fig. 36. Principal circuit of the "vibrometer method."

which is Eq. (58b) for $V_2 = 0$. The high-level characteristics of the force factors, A_1 and A_2, are measurable as described later. The force factors depend only on the vibrational velocity v. Therefore, the velocity can be obtained uniquely from the measured value of I_m, which is designated as motional current.

With no external mechanical force, i.e., $F = 0$, the total mechanical impedance of the vibrometer system inclusive of the radiation impedance follows from Eq. (58):

$$z_a + z_0 = -\frac{A_1 A_2}{I_m} V_1 \tag{60}$$

Therefore, when the motional current I_m is kept constant, the resonant curve of the driving voltage V_1 represents the frequency characteristics of the total mechanical impedance $|z_a + z_0|$ for a given amplitude of the vibrational velocity v. This relation holds even in the case of large amplitude vibrations as long as every quantity is observed for its fundamental value. Thus, the quadrantal frequency-difference Δf of the mechanical resonance of the vibrometer system can be obtained from the measured resonance curve of the driving voltage V_1. The total mechanical resistance $(r_a + r_0)$ can be expressed as

$$r_a + r_0 = 2\pi m \Delta f \tag{61}$$

where m is the effective mass of the vibration system.

7.2.3. Differential Method

The principle of the differential method is shown in Fig. 37. Two vibrators with originally identical characteristics are used. One of the vibrators, S_1, operates in the acoustic medium at its mechanical resonance. On the surfaces of the other vibrator, S_2, however, suitable blocks of metal are cemented to shift the mechanical resonance. Then, the S_2-vibrator is in a quasi-clamped state during measurement. The high-frequency source drives both vibrators in parallel. A differential transformer, T, is used to pick up the difference between the individual currents flowing in each vibrator. Because the clamped admittances of both vibrators are the same, the difference current is identical with the motional current of the S_1-vibrator in this case

$$I_m = I_1 - I_2 \tag{62}$$

Using a set of fundamental electroacoustic equations, the motional

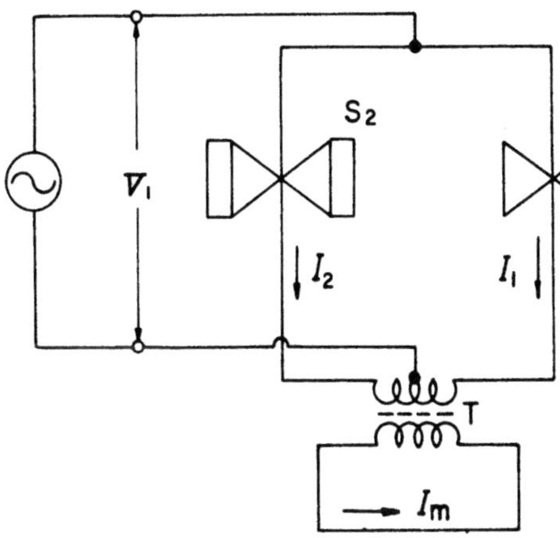

Fig. 37. Principal circuit of the "differential method."

current can be expressed as

$$I_m = -A_1 v \tag{63}$$

and the total mechanical impedance of the S_1-vibrator as

$$z_a + z_0 = -\frac{A_1^2}{I_m} V_1 \tag{64}$$

where A_1 is the force factor of the S_1-vibrator. It is easily seen that these equations are the same as those for the vibrometer method described in the foregoing section. The same measuring arrangement can be used also in this case.

7.2.4. High-Level Characteristics of Force Factor

Denoting the driving voltage V_1 at the resonant frequency by V_{10}, the total mechanical resistance r can be written from Eq. (60) as follows

$$r = -\frac{A_1 A_2}{I_m} V_{10} \tag{65}$$

Then

$$|A_1 \cdot A_2| = r\left|\frac{I_m}{V_{10}}\right| \tag{66}$$

where I_m/V_{10} is an admittance, called the motional admittance at resonance. Denoting it by Y_{mo}, we have

$$Y_{mo} = |I_m/V_{10}| \tag{67}$$

Eq. (66), therefore, will be

$$| A_1 \cdot A_2 | = 2\pi m \, \Delta f | I_m / V_{10} |$$

or

$$| A_1 \cdot A_2 | = 2\pi m \, \Delta f | Y_{mo} | \tag{68}$$

Thus, the value of $| A_1 \cdot A_2 |$ at any high level of vibration can be obtained by one measurement of either V_{10} or Y_{mo}; Y_{mo} can be read directly.[102]

Figure 38 shows an example for the observed results of the force-factor characteristics when the vibrometer is made of magnetostrictive ferrite (Ni–Cu–Co ferrite). The force factor has a tendency to decrease slowly to some extent when the vibrational velocity increases to a higher level. In the figure, the circles show measurements carried out in a liquid medium and the plots of crosses show those in air; it should be noted that the characteristics of the force factors are independent of the change of acoustic load, but dependent only on the vibrational velocity. The separation of A_1 and A_2 is made under an assumption that the ratio of A_1 to A_2 is always

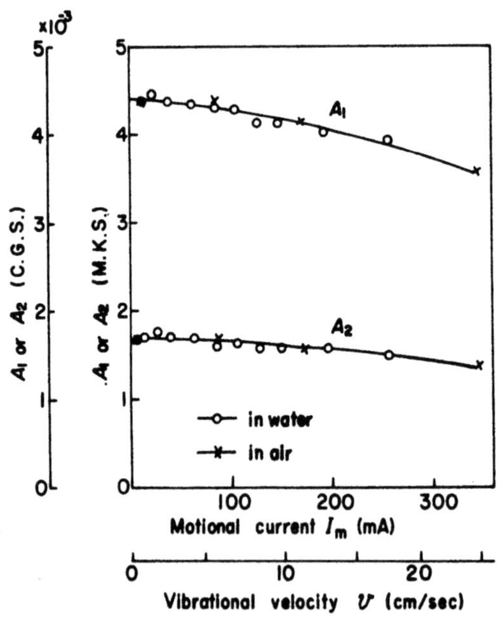

Fig. 38. An example of the high-level characteristics of the force factor observed in a magnetostrictive ferrite vibrator. Individual characteristics of the force factor A_1 and A_2 are obtained by a separation analysis. The additional scale below is the corresponding vibrational velocity ν.

constant even at a high level of vibration. These data are for a biasing magnetization, which is nearly one half of the saturation magnetization. The decreasing tendency of the force factor at higher levels can be greatly improved when the biasing magnetization is increased.[103]

7.2.5. Other Methods of Approach

Golyamina and Chulkova[104] reported a method for measuring the dependence of the mechanical Q, Young's modulus, and magnetostrictive constant on the amplitudes of alternating stress and magnetic induction. Some interesting data were disclosed with regard to various magnetostrictive ferrites, stating that: (1) Young's modulus E^H and the mechanical Q^H at an optimum biasing magnetization decrease with the increase of alternating mechanical stress in a rod-type vibrator. The amount of the decrease, however, is very much dependent on the compositions of the ferrites. (2) The magnetostrictive driving force induced by an alternating magnetization has an attainable ultimate value in some ferrites. The value, however, varies when the biasing field is varied. But, whatever the value may be, with loading in air or in water under cavitation, the limiting output power is completely determined by the mechanical strength of the material.

REFERENCES

1. L. Balamuth, U.S. Patent No. 2, 580, 716 (Jan. 1, 1952); *IRE Convention Record*, Vol. 3, Part 9, "Ultrasonics II," 89–97 (1955).
2. T. F. Hueter and R. H. Bolt, *Sonics*, John Wiley & Sons, Inc., New York (1955), pp. 276–277.
3. J. B. Jones and J. J. Powers, "Ultrasonic Welding," *The Welding Journal*, 761–766 (August, 1956).
4. J. B. Jones and F. R. Meyer, "Ultrasonic welding of structural aluminium alloys," *Supplement to the Welding Journal*, 81s–92s (March, 1958).
5a. E. Mori, M. Inoue, and S. Toda, "The experiment of ultrasonic wire drawing," (in Japanese), *Reports of the 1965 Spring Meeting, The Acoustical Society of Japan, No.* 2-1-20 (May, 1965).
5b. J. R. Frederick, *Ultrasonic Engineering*, John Wiley & Sons, Inc., New York (1965), pp. 169–170.
6. Y. Kikuchi and K. Fukushima, "The performance theory and the design procedure of the laminated magnetostriction vibrators," (in English), *Sci. Rept. Res. Inst., Tohoku Univ.*, **B-1-2**(1), 141–189 (1951).
7. Y. Kikuchi, in *Ultrasonic Transducers* (in English) (Y. Kikuchi, ed.), Corona Publishing Co. Ltd., Tokyo (1969), Chap. 5.
8. Y. Kikuchi and H. Shimizu, "Magnetostrictive ultrasonic transducers made of ferrites," (in English), *Sci. Rept. Res. Inst., Tohoku Univ.* **B-7**(1), 9–15 (1955).

9. Y. Kikuchi, "Performance of magnetostrictive transducers made of aluminum–iron alloy or nickel–copper ferrite," *J. Acoust. Soc. Am.* **29**(5), 569–573 (1957).
10. Y. Kikuchi, H. Shimizu, and D. Okuyama, "Fork-type magnetostriction vibrators made of ferrites," (in Japanese), *J. Inst. Elec. Commun. Eng. Japan*, Paper prepared for the 1960 Joint Meeting in Sapporo, No. 1075, (July 1960).
11. H. H. Hall, "A magnetostriction filter," *Proc. IRE* **21**(9), 1328–1338 (1933).
12. K. Fukushima and T. Koitibara, "Differential magnetostriction filter," (in Japanese), *J. Inst. Elec. Commun. Engr. (Japan)*, **194**, 233–238 (1939).
13. A. P. Thiele, "Narrow-band magnetostrictive filters," *Electronics and Radio Engineer*, **35**(11), 402–411 (1958).
14. K. Shibayama, Y. Kikuchi and T. Sato, "A new ferrite magnetostrictive resonator of composed type for magnetostriction filters," (in Japanese), *Reports of the 1965 Spring Meeting, The Acoustical Society of Japan*, No. 1-1-9 (May, 1965).
15. Y. Kikuchi, "Magnetostrictive materials and applications," *IEEE Transactions on Magnetics* **Mag-4**(2), 107–117 (1968).
16. K. Shibayama, in *Ultrasonic Transducers* (in English) (Y. Kikuchi, ed.), Corona Publishing Co. Ltd., Tokyo (1969), Chap. 11.
17. R. Boynton, "A magnetostrictive-filter random wave analyzer," *IRE International Convention Record*, Vol. 8, Pt. 9 (1960), pp. 217–226.
18. A. B. Wood, "A Textbook of Sound," G. Bell and Sons Ltd., London (1930), pp. 154–155.
19. H. Nukiyama and K. Aoyagi, Japanese Patent 109, 289 (1935).
20. A. B. Wood, F. D. Smith, and J. A. McGeachy, "A magnetostriction echo depth-recorder," *J. Inst. Elec. Engr.* **76**, 550–566 (1935).
21. Y. Kikuchi, in *Ultrasonic Transducers* (in English) (Y. Kikuchi, ed.), Corona Publishing Co. Ltd., Tokyo (1969), Chap. 1.
22. Y. Kikuchi, in *Ultrasonic Transducers* (in English) (Y. Kikuchi, ed.), Corona Publishing Co. Ltd., Tokyo (1969), Chap. 3.
23. Y. Kikuchi, "On the magnetostrictivity" (in English), *Nippon Elec. Commun. Eng.*, No. 30 (October, 1942).
24. F. D. Smith, "The magnetostriction constant for alternating magnetic field," *Proc. Phys. Soc. (London)* **42**, 181–191 (1930).
25. S. Butterworth and F. D. Smith, "The equivalent circuit of the magnetostriction oscillator," *Proc. Phys. Soc. (London)* **43**, 166–185 (1931).
26. H. Sussman and S. L. Ehrlich, "Evaluation of the magnetostrictive properties of Hiperco," *J. Acoust. Soc. Am.* **22**, 499–506 (1950).
27. C. M. van der Burgt, "Performance of ceramic ferrite resonators as transducers and filter elements," *J. Acoust. Soc. Am.*, **28**, 1020–1032 (1956).
28. F. V. Hunt, *Electroacoustics*, John Wiley & Sons, Inc., New York (1954), p. 129.
29. Y. Kikuchi and H. Shimizu, "On the effective attenuation of some electroacoustic transducers at their conjugate electrical matching," (in English), *Sci. Rept. Res. Inst., Tohoku Univ.*, **B-3**(1), 13–18 (1951).
30. Y. Kikuchi and H. Shimizu, "On the effective attenuation of ring-type magnetostriction transducer," (in English), *Sci. Rept. Res. Inst., Tohoku Univ.* **B-3**(1), 1–5 (1951).
31. Y. Kikuchi and H. Shimizu, "On the B-type resonance of magnetostriction and electrostriction transducers," (in English), *Sci. Rept. Res. Inst., Tohoku Univ.* **B-4**(1), 173–203 (1952).

32. Y. Kikuchi, *Magnetostriction Vibration and its Application to Ultrasonics*, (in Japanese), Corona Publishing Co. Ltd., Tokyo, 5th Edition (1966).
33. R. M. Bozorth, *Ferromagnetism*, Van Nostrand Co., New York (1951), p. 691.
34. Y. Kikuchi and H. Shimizu, "On the relation between ΔE-effect and electromechanical coupling factor in magnetostrictive materials," (in Japanese), *J. Inst. Elec. Commun. Engr. Japan*, No. 717 (April, 1959). Paper prepared for the 1959 Joint Meeting.
35. Y. Kikuchi, in *Ultrasonic Transducers*, (in English) (Y. Kikuchi, ed.), Corona Publishing Co. Ltd., Tokyo (1969), Chap. 3.
36. R. M. Bozorth, *Ferromagnetism*, Van Nostrand Co., New York (1951), p. 545.
37. Henry B. Karplus, "Effect of hydrostatic pressure on magnetostrictors," *J. Acoust. Soc. Am.* **35**, 800 (1963).
38. Y. Kikuchi, H. Shimizu, and T. Uekusa, "Force factor of magnetostriction vibrators under hydrostatic pressure of 70 atm," (in Japanese), Reports of the 1964 Spring Meeting, *The Acoustical Soc. of Japan*, No. 1-1-8 (May, 1964).
39. Y. Kikuchi, H. Shimizu, and T. Uekusa, "Force factor of magnetostriction vibrators under hydrostatic pressure of 70 atm," (in Japanese), Record Elec. Commun. Eng. Conversazione, Tohoku Univ., Vol. 34, No. 1 (1965), p. 46.
40. R. M. Bozorth, *Ferromagnetism*, Van Nostrand Co., NewYork (1951), p.634.
41. Y. Kikuchi and H. Shimizu, "Formulated expression of magnetostriction characteristics under mechanical stress and the consideration on magnetostrictive vibration of large amplitude," (in Japanese), Record of Ultrasonic Professional Group, The Institution of Electric and Communication Engineers of Japan (Denki-Tsushin-Gakkai, Shiba, Minato-ward, Tokyo), (May, 1961).
42. Y. Kikuchi, *Ultrasonic Transducers*, (in English) (Y. Kikuchi, ed.), Corona Publishing Co. Ltd., Tokyo (1969), Chap. 4, p. 104.
43. Y. Kikuchi, *Magnetostriction Vibration and its Application to Ultrasonics*, (in Japanese), Corona Publishing Co. Ltd., Tokyo, 5th Edition (1966), p. 80.
44. C. M. Davis and S. F. Ferebee, "Dynamic magnetostrictive properties of Alfenol," *J. Acoust. Soc. Am.* **28**(2), 286–290 (1956).
45. Y. Kikuchi, *Ultrasonic Transducers*, (in English) (Y. Kikuchi, ed.), Corona Publishing Co. Ltd., Tokyo (1969), Chap. 4.
46. Y. Kikuchi and K. Fukushima, *J. Inst. Telegraph and Telephone Engr. Japan*, 382 (1942).
47. K. Fukushima and A. Kurobe, "Improvement of magnetostrictivity measurement by means of mutual motional impedance method," (Tentative translation from the Japanese title), *Record Elec. Commun. Eng. Conversazione, Tohoku Univ.* (1948).
48. Y. Kikuchi, H. Shimizu, and Y. Sanpei, "Improvement of magnetostrictivity meter," (in Japanese), Paper prepared for Acoustics Conversazione, Research Institute of Electrical Communication, Tohoku University, (September 17, 1954); and or Paper prepared for the Meeting of Acoustical Society of Japan (May, 1956).
49. Y. Kikuchi, H. Shimizu, and M. Terashima, "A new method for precise measurement of quadrantal frequency difference by applying carrier-suppressed modulation," (in English), *Sci. Rept. Res. Inst., Tohoku Univ.* **B-7**(1), 17–21 (1955).
50. Y. Kikuchi, H. Shimizu, and T. Uekusa, "Sound reflection coefficient of wedge-type absorbers made of pine under high hydrostatic pressure," *Record Elec. Commun. Eng. Conversazione, Tohoku Univ.* **34**(1), 48 (1965).

51. T. Uekusa, "On the radiation of intense ultrasonic waves under large hydrostatic pressure," (Tentative translation of the title of the original Japanese text), Doctoral Thesis, Tohoku University, Sendai, Japan (1966).
52. R. R. Whymark and James M. Witting, "Resonant displacement of nickel and permendur magnetostrictors under static compressive stress," *J. Acoust. Soc. Am.* **33**, 1720 (1961).
53. W. Döring, "Über die Temperaturabhängigkeit der Magnetostriktion von Nickel, *Z. Physik* **103**, 560–582 (1936).
54. H. Kirchner, Über den Einfluss von Zug, Druck, und Torsion auf die Längsmagnetostriktion, *Ann. Physik* **27**, 49–69 (1936).
55. Y. Kikuchi and K. Fukushima, "On the magnetostriction activity in highly reduced plates of nickel," (in Japanese), *J. Inst. Elec. Engrs. Japan* **62**(647), 325 (1942).
56. Y. Kikuchi, "Empirical formula for magnetostriction phenomena in nickel," (in English), *Sci. Rept. Res. Inst., Tohoku Univ.* **B-4**(2), 255–258 (1953). (The original Japanese publication was in 1942.)
57. E. R. W. Jones, C. A. Clark, and E. A. Fell, "Calculated and observed effects of texture on the magnetic properties and Young's modulus of nickel sheet," *Brit. J. Appl. Phys.* **9**(5), 178–184 (1958).
58. Y. Mashiyama, "On the hysteresis of magnetostriction of iron, nickel, cobalt, and single crystals of iron," (in English), *Sci. Rept. Tohoku Univ.* **26**, 1–39 (1937).
59. Y. Kikuchi, see Ref. (23).
60. Y. Kikuchi and H. Shimizu, see Ref. (41).
61. Y. Kikuchi and H. Shimizu, "Effect of oxide-forming heat treatment on the magnetic and magnetostrictive characteristics of nickel," (in Japanese), Record of 1954 Joint Meeting of Three Institutes of Electrical Engineers of Japan in Tokyo, No. 204 (October, 1954).
62. A. Shulze, "Die Magnetostriktion. I," *Z. Physik* **50**, 448–505 (1928).
63. Y. Nishina, K. Fukushima, and H. Inoue, "On the magnetostriction vibration of iron–nickel alloy," (Tentative translation from the Japanese title), *J. Inst. Elec. Commun. Engr. Japan*, 24–25 (October, 1937).
64. C. M. Davis, Jr., H. H. Helms, and S. F. Ferebee, "Dynamic magnetostrictive properties of Ni–Fe alloys," *J. Acoust. Soc. Am.* **29**(4), 431–434 (1957).
65. K. Honda, H. Masumoto, Y. Shirakawa and T. Kobayashi, "On the magnetostriction of iron–aluminum alloys and a new alloy 'Alfer'," Oral presentation in 1941 at the Sendai Meeting of the Japan Institute of Metals; *Sci. Rept. Res. Inst., Tohoku Univ.* **A-1**(4), 341–347 (1949).
66. H. Masumoto and G. Otomo, "On the dynamical characteristics of the magnetostriction alloy Alfer," (in English), *Sci. Rept. Res. Inst. Tohoku Univ.*, **A-2**(3), 413–419 (1950).
67. J. F. Nachman and W. J. Buehler, 16% aluminum–iron alloy cold rolled in the order–disorder temperature range, *J. Appl. Phys.* **25**(3), 307–313 (1954).
68. C. M. Davis and S. F. Ferebee, "Dynamic magnetostrictive properties of Alfenol," *J. Acoust. Soc. Am.* **28**(2), 286–290 (1956).
69. M. T. Pigott and P. M. Kendig, "Iron–aluminum alloys for use in magnetostrictive transducers," *J. Acoust. Soc. Am.* **28**(3), 343–346 (1956).
70. O. Henkel, Eisen–Aluminium-Legierungen für magnetostriktive Wandler, *Hochfrequenztechnik und Elektroakustik* **71**(5), 161–166 (1962).
71. Z. N. Bulycheva, E. I. Gurvich, and Ya. P. Selisskii, "Magnetic alloys used in ultrasonics," *Soviet Progress in Applied Ultrasonics*, Vol. 1 (1964), pp. 190–196.

72. O. N. Al'tgauzen, L. S. Bezuglaya, Z. N. Bulycheva, and O. V. Lyubetskaya, "Magnetic properties of alloys for magnetostrictive transducers," *Soviet Phys.–Acoust.* **12**(3), 249–253 (1967).
73. Z. N. Bulycheva, M. M. Borodkina, and V. L. Sandomirskaya, "Texture and magnetic properties of Fe–Al magnetostriction alloys," *The Physics of Metals and Metallography* **19**(1), 147–150 (1965).
74. R. C. Hall, "Magnetostriction of aluminum–iron single crystals in the region of 6 to 30 atom% aluminum," *J. Appl. Phys.* **28**(6), 707–713 (1957).
75. H. Gengnagel, "Das magnetische Verhalten von Einkristallen aus Fe–Al-Legierungen im Überstrukturgebiet Fe_3Al. II. Magnetostriktion und spez. Widerstand," *Naturwiss.* **45**(4), 81–82 (1958).
76. K. Fukushima, H. Nukiyama, and Y. Nishina, private communication.
77. Y. Kikuchi and M. Nishikawa, "Magnetostrictive characteristics of magnetically anisotropic silicon steel sheet," (in Japanese), *J. Inst. Elec. Engr.* (*Japan*) **63**(659), 458 (1943).
78. C. A. Clark, "The dynamic magnetostriction of nickel–cobalt alloys, *Brit. J. Appl. Phys.* **7**(10), 355–360 (1956).
79. W. T. Harris, K. H. Farr, C. A. Clark, and E. M. Wise, "Magnetostrictive properties of certain cobalt–nickel alloys, *J. Acoust. Soc. Am.* **31**, 854 (1959).
80. C. A. Clark, "Improved nickel-base alloys for magnetostrictive transducers," *J. Acoust. Soc. Am.* **33**, 930–933 (1961).
81. C. A. Clark and J. J. Mason, "Temperature dependence of dynamic magnetostrictive properties in nickel–cobalt–chromium alloys, *J. Acoust. Soc. Am.* **35**, 1665 (1963).
82. Z. N. Bulycheva, E. I. Gurvich, and Ya. P. Selisskii, "Magnetic alloys used in ultrasonics," *Soviet Progress in Applied Ultrasonics*, **1**, 190–196 (1964).
83. H. E. Strauss, "Magnetostriction of low-density cobalt rondel," *J. Appl. Phys.* **29**(12), 1690–1691 (1958).
84. T. Takei and H. Nakamura, "On the magnetostriction of metal-oxide magnetic substances," (Tentative translation of the title of the original Japanese text), Paper prepared for the 22nd Joint Meeting of the Institute of Electrical Engineers (Japan), No. 36 (April, 1943), p. 45.
85. C. M. van der Burgt, "Dynamical physical parameters of the magnetostrictive excitation of extensional and torsional vibrations in ferrites," *Philips Research Reports* **8**, 91–132 (1953).
86. H. Thiede, "Untersuchungen an Ferriten auf ihre Eignung als Flüssigkeitsschallwandler, *Acustica* **4**, 532–536 (1954).
87. Y. Kikuchi, N. Tsuya, H. Shimizu, M. Terajima, A. Sugiyama, T. Hirone, S. Maeda, and J. Shimoiizaka," Study on ferrite for the use in magnetostriction vibrator, Part I, Ni–Zn ferrite," (in English), *Sci. Rept. Res. Inst. Tohoku Univ.* **B-7**(1), 1–7 (1955).
88. Y. Kikuchi, N. Tsuya, H. Shimizu, M. Terajima, A. Sugiyama, T. Hirone, S. Maeda, and J. Shimoiizaka, "Study on ferrite for the use in magnetostriction vibrator, Part II, Ni–Cu ferrite," (in English), *Sci. Rept. Res. Inst., Tohoku Univ.* **B-7**(3), 171–178 (1955).
89. Y. Kikuchi, K. Fukushima, N. Tsuya, H. Shimizu, T. Hirone, S. Maeda, J. Shimoiizaka, and E. Fujino, "Dynamic magnetostriction characteristics of Ni–Cu–Co ferrites," (in Japanese), Paper prepared for 1957 National Meeting of the Institute of Electronics and Communication Engineers of Japan, No. 39 (November, 1957).

90. H. Shimizu and T. Chiba, "Dynamic magnetostriction characteristics of Ni–Cu–Co–Zn ferrites," (Tentative translation of the original Japanese print), Paper prepared for Acoustics Conversazione, Research Institute of Electrical Communication, Tohoku University (March, 1960).
91. C. M. van der Burgt, Ferrox-cube material for piezomagnetic vibrators, *Philips Tech. Rev.* **18**(10), 285–298 (1956/57).
92. Private communication from Tohoku Metal Industries Ltd., Sendai, Japan.
93. S. F. Ferebee and C. M. Davis, Jr., "Effect of divalent ion substitutions on the magnetomechanical properties of nickel ferrite," *J. Acoust. Soc. Am.* **30**(8), 747–750 (1958).
94. A. D. Sokolov and Ya S. Shur, "Connection between the magnetic properties and sensitivity of magnetostrictive nickel–zinc ferrite pickups," Soviet Phys.–Acoust. **6**(1), 130–132 (1960).
95. I. P. Golyamina, "Magnetostrictive ferrites as a material for electroacoustic transducers," *Soviet Phys.–Acoust.* **6**(3), 311–320 (1960).
96. G. Sherwood Smith and Oskar Mattiat, "Controlled temperature dependence of magnetostrictive ferrites," *J. Acoust. Soc. Am.* **32**, 1499 (1960).
97. C. M. van der Burgt and A. L. Stuijts, "Low porosity ferrites for high-intensity ultrasonic radiators," The 4th International Congress on Acoustics K22, Copenhagen (August, 1962).
98. Z. Koczkowski, "Hysteresis of the magnetomechanical parameters of type El magnetostrictive ferrite," *Soviet Phys.–Acoust.* **9**(1), 29–36 (1963).
99. I. P. Golyamina, "Application of ferrite as ultrasonic transmitters, *Soviet Progress in Applied Ultrasonics* **1**, 183–189 (1964).
100. L. I. Ganeva and I. P. Golyamina, "Properties of magnetostrictive ferrites at high temperatures," *Soviet Phys.–Acoust.* **9**(4), 337–340 (1964).
101. Y. Kikuchi and T. Matsuno, "Properties of ferrite magnetostrictive vibrators at high temperatures," (in Japanese), Reports of the 1965 Spring Meeting, The Acoustical Society of Japan, No. 1-1-7 (May, 1965).
102. Y. Kikuchi, H. Shimizu and D. Okuyama, "Method for measuring acoustic radiation resistance of a transducer generating intense ultrasonic cavitation," (in English), *Sci. Rept. Res. Inst., Tohoku Univ.* **B-17** (3–4), 51–63 (1965). Original Japanese publication; Measurement of acoustic-radiation-resistance variation due to cavitation, *Record Elec. Commun. Eng. Conversazione, Tohoku Univ.* **27**(2), 33–34 (1958).
103. Y. Kikuchi, H. Shimizu, T. Sannomiya, and T. Uekusa, "The effect of bias magnetic field on the large amplitude characteristics of ferrite magnetostriction vibrator," (in Japanese), Reports of the 1966 Autumn Meeting, The Acoustical Society of Japan, No. 1-1-11 (November, 1966).
104. I. P. Golyamina and V. K. Chulkova, "Properties of magnetostrictive ferrites at large amplitudes of magnetic induction and mechanical stress," *Soviet Phys.–Acoust.* **10**(3), 236–241 (1965).
105. Y. Kikuchi, On the synthesis of the theoretical characteristics of magnetostriction in polycrystal metals," (in English), *Sci. Rept. Res. Inst., Tohoku Univ.* **B-4**(2), 259–286 (1953). Original Japanese publication: *J. Inst. Elec. Engrs.* (*Japan*) **63** (660), 471–473 (1943).
106. W. Heisenberg, "Zur Theorie der Magnetostriktion und der Magnetisierungskurve," *Z. Physik.* **69**, 287–297 (1931).
107. R. Gans and J. von Harlem, "Magnetostriktion ferromagnetischer Kristalle," *Ann. Physik.* **16**, 162–173 (1933).

108. K. Honda and Y. Mashiyama, "On the magnetostriction of single crystals of iron," (in English), *Sci. Rept. Tohoku Univ.* **15**, 755–776 (1926).
109. W. L. Webster, "Magnetostriction in iron crystals," *Proc. Roy. Soc.* (*London*), (*A*)**109**, 570–584 (1925).
110. H. B. Callen and N. Goldberg, "Magnetostriction of polycrystalline aggregates," *J. Appl. Phys.* **36**(3–2), 976–977 (1965).
111. R. Kimura and K. Ohno, "On the elastic constants of single crystals of iron," *Sci. Rept. Tohoku Univ.* **23**, 359–364 (1934).
112. K. Honda and Y. Shirakawa, "On Young's modulus of elasticity of single crystal of nickel," *J. Japan Inst. Metals* **1**, 217–219 (1937).
113. M. Yamamoto, "On the ΔE-effect of iron, nickel, and cobalt," *J. Japan Inst. Metals* **5**, 167–174 (1941).
114. E. J. Langham, "Magnetostriction of polycrystalline silicon–iron sheet," *Brit. J. Appl. Phys.* **14**, 16–19 (1963).
115. Y. Kikuchi, "Theoretical investigation on the ultimate mechanical output power available from magnetostriction vibrators," (in Japanese), *J. Inst. Elec. Commun. Engr. Japan* **29**, 197–208 (1946).
116. Y. Kikuchi and H. Shimizu, "Theoretical investigation on the ultimate mechanical output power by magnetostriction vibration," (in Japanese), Record of Annual Meeting, The Acoustical Society of Japan, No. 1-3-17 (May, 1957).
117. Y. Kikuchi, "The theoretical limit of mechanical output power available from magnetostriction vibrators," *IEEE Transactions on Sonics and Ultrasonics* **SU-15**, 1–13 (1968).
118. Y. Kikuchi and H. Shimizu, "On the variation of acoustic radiation resistance in water under ultrasonic cavitation," *J. Acoust. Soc. Am.* **31**(10), 1385–1386 (1959).
119. Y. Kikuchi, H. Shimizu, and D. Okuyama, in *Ultrasonic Energy* (E. Kelly, ed.) University of Illinois Press, Urbana (1965), pp. 23–50.
120. Y. Kikuchi, H. Shimizu, and D. Okuyama, "Some aspects of ultrasonic cavitation; *Sci. Rept. Res. Inst., Tohoku Univ.* **B-17**(3–4), 65–85 (1965).
121. K. Tsukamoto, "On fatigue phenomena in the magnetostriction characteristics of nickel after intense vibration," (in Japanese), Graduate Thesis, Faculty of Engineering, Tohoku University (March 15, 1950).
122. Y. Kikuchi and H. Shimizu, "On the magnetic hysteresis losses in magnetostriction vibration," (in English), *Sci. Rept. Res. Inst., Tohoku Univ.* **B-1-2**(3), 365–379 (1951).
123. R. R. Whymark, "Intense sound output of a nickel magnetostrictive transducer," *J. Acoust. Soc. Am.* **31**, 1570 (1959).
124. R. R. Whymark, "Utilization of magnetostrictive materials in generating intense sound," *J. Acoust. Soc. Am.* **33**, 725 (1961).
125. C. M. van der Burgt, "Piezomagnetic Ferrites," *Electronic Technology*, 330–341, September (1960).

Chapter 2

PIEZOELECTRIC CRYSTALS AND CERAMICS

Don Berlincourt

Vernitron Piezoelectric Division
Bedford, Ohio

1. INTRODUCTION*

Piezoelectric crystals and ceramics are used as detectors and radiators of acoustic power from very low frequencies to above 10^9 Hz. The lower frequency range is now covered almost exclusively by the piezoelectric ceramics which date from the late 1940's, and the upper frequency range is covered by new piezoelectric crystals discovered in the 1960's. The very highest frequency range has been opened up by oriented deposited films of several of the new crystals.

Emphasis in this chapter is on the newer developments, primarily those since about 1955. This period has seen the introduction of the lead titanate zirconates which have become the dominant transducer material in the range to a few MHz, covering transducers for sonar and various ultrasonic applications. Also during this period a number of important piezoelectric crystals were discovered, in most cases in connection with the search for crystals with interesting semiconductor or electrooptical characteristics. Improved high-temperature crystal growth techniques played a major part in these developments.

Nomenclature and units used in this chapter are those in the *IRE Standards on Piezoelectric Crystals.*[1–3] More complete and rigorous developments of the piezoelectric relations and more complete data on the older piezoelectric crystals may be found in several books and articles.[4–11]

* See p. 119 for a list of notation used in this chapter.

The Landolt–Börnstein volume is the most complete source for electroelastic constants and their temperature variation.

2. FUNDAMENTALS OF PIEZOELECTRICITY

2.1. Basic Action and Linear Static Equations

The piezoelectric equations relate mechanical and electric variables. For practical purposes the effects of magnetic fields are ignored when dealing with piezoelectric materials, and effects of electric fields are ignored when dealing with piezomagnetic materials. Static conditions are isothermal, but for vibrating elements adiabatic conditions usually prevail. For practical purposes, however, differences between adiabatic and isothermal coefficients can be ignored.*

There are four forms of the linear piezoelectric equations, the form depending on the choice of electric field (E) or charge density (D) on the one hand or mechanical stress (T) or strain (S) on the other as the independent variables. The more useful forms are those with E and S as the independent variables for thickness modes, and E and T as the independent variables for longitudinal modes with transverse E. The sets of piezoelectric constants are e_{ijk} and d_{ijk} respectively.

Ignoring thermal and magnetic effects these equations are listed below in tensor form.† Note that matrix rather than tensor notation is used for simplicity.‡

$$\begin{aligned} T_i &= c_{ij}^E S_j - e_{mi} E_m \\ D_m &= e_{mi} S_i + \varepsilon_{mk}^S E_k \end{aligned} \tag{1}$$

$$\begin{aligned} S_i &= s_{ij}^E T_j + d_{mi} E_m \\ D_m &= d_{mi} T_i + \varepsilon_{mk}^T E_k \end{aligned} \tag{2}$$

Here a repeated suffix denotes a summation with respect to the suffix, with $i, j = 1$ to 6 and $m, k = 1$ to 3.

The only other useful set of piezoelectric equations, for longitudinal modes with parallel electric field, is that with stress and charge density as the independent variables (Eq. (3)). For completeness, however, the set

* See p. 187 of Ref. (7).

† See p. 182 of Ref. (7) for the development from the thermodynamic potentials.

‡ $T_{ii} = T_i$, $T_{23} = T_4$, $T_{13} = T_5$, $T_{12} = T_6$, $S_{ii} = S_i$, $2S_{12} = S_6$, $2S_{23} = S_4$, $2S_{13} = S_5$.

with strain and charge density as the independent variables is also listed (Eq. (4)).

$$\begin{aligned} S_i &= s_{ij}^D T_j + g_{mi} D_m \\ E_m &= -g_{mi} T_i + \beta_{mk}^T D_k \end{aligned} \tag{3}$$

$$\begin{aligned} T_i &= c_{ij}^D S_j - h_{mi} D_m \\ E_m &= -h_{mi} S_i + \beta_{mk}^S D_k \end{aligned} \tag{4}$$

The piezoelectric constants are third-order tensors relating second-order symmetric tensors (strain or stress) to vectors (charge density or electric field). The elastic constants s_{ij} and c_{ij} are fourth-order symmetric tensors relating two second-order symmetric tensors. The dielectric constants ε_{mk} and β_{mk} are second-order symmetric tensors relating two vectors.

The constants g and h vary much less from one piezoelectric material to another than do d and e. The constants g and h also do not show the extreme temperature dependence found for d and e with ferroelectric materials.

The elastic compliances s_{ij} and dielectric impermeabilities β_{mk} are related to the elastic stiffness constants c_{ij} and dielectric permittivities ε_{mk} respectively as follows:

$$s_{ij} = \frac{(-1)^{i+j} \Delta_{ij}}{\Delta}, \tag{5a}$$

$$\beta_{mk} = \frac{(-1)^{m+k} \Delta_{mk}}{\Delta}. \tag{5b}$$

In Eq. (5a) Δ is the determinant of the c_{ij} matrix and Δ_{ij} is the minor obtained by excluding the ith row and jth column. In Eq. (5b) Δ is the determinant of the ε_{mk} matrix and Δ_{mk} is its minor. Identical relationships hold for obtaining c_{ij} from the s_{ij} matrix and ε_{mk} from the β_{mk} matrix.

The definitions of the piezoelectric constants follow directly from Eqs. (1)–(4), for instance,

$$e = \left(-\frac{\partial T}{\partial E}\right)_S = \left(\frac{\partial D}{\partial S}\right)_E \tag{6}$$

The piezoelectric constants of Eqs. (1)–(4) are interrelated as follows:

$$\begin{aligned} d_{mi} &= \varepsilon_{nm}^T g_{ni} = e_{mj} s_{ji}^E \\ g_{mi} &= \beta_{nm}^T d_{ni} = h_{mj} s_{ji}^D \\ e_{mi} &= \varepsilon_{nm}^S h_{ni} = d_{mj} c_{ji}^E \\ h_{mi} &= \beta_{nm}^S e_{ni} = g_{mj} c_{ji}^D \end{aligned} \tag{7}$$

2.2. Effect of Crystal Symmetry

Since lack of a center of symmetry is a necessary condition for the occurrence of piezoelectricity, all piezoelectric materials are anisotropic. In the asymmetric class there are 18 independent piezoelectric constants, 21 elastic constants, and 6 dielectric constants. Due to the piezoelectric interaction there are two sets of dielectric and elastic constants, constant strain ε^S and constant stress ε^T for the dielectric constants, and constant electric field c^E and constant charge density c^D for elastic constants.

With increasing symmetry the number of independent constants decreases drastically and many coefficients disappear. Technically important piezoelectric crystals and the piezoelectric ceramics have relatively high symmetry. The reduction in the number of independent constants by symmetry is treated in the books by Cady,[4] Mason,[5] and Nye.[12] Cady thoroughly discusses methods to obtain effective electroelastic constants for rotated cuts in terms of those for the orthogonal axes. The convention for choice of a set of orthogonal (X_1, X_2, X_3) axes in terms of the crystallographic axes is discussed in the IRE Standards.[1] The 1958 Standard[2] shows complete electroelastic matrices for the 32 crystal classes, 20 of which are piezoelectric. Table I shows the complete matrices for piezoelectric materials discussed in this book.

2.3. The Piezoelectric Coupling Factor

The coupling factors are among the most important properties of piezoelectric materials. For our purposes the coupling factor may be considered to be the square root of the ratio of electrical (mechanical) work which can be done under ideal conditions to the total energy stored from a mechanical (electrical) source. The piezoelectric coupling factors of piezoelectric materials of much different levels of permittivity and compliance may be directly compared. In this section the material coupling factor k_{mat} is considered. This differs in the general case from the effective coupling factor discussed in Section 3.3 because in most cases there is a spatial distribution of elastic and dielectric energy in resonant vibration.

The material coupling factors occur as natural combinations of electroelastic constants. For instance, consider a piezoelectric element plated on the faces perpendicular to x_3 and short-circuited, to which a stress T_3 is applied (Fig. 1a). Consider also that the element is free to cross expand freely, so that T_3 is the only component of stress. The total stored energy per unit volume at maximum compression is given by $\Sigma \int T_i \, dS_i = \int T_3 \, dS_3$.

Table I. Electroelastic Matrices for Common Piezoelectric Classes

Class 2

(*Rochelle salt in ferroelectric range, lithium sulfate*)

$$\begin{vmatrix} s_{11} & s_{12} & s_{13} & 0 & s_{15} & 0 \\ s_{12} & s_{22} & s_{23} & 0 & s_{25} & 0 \\ s_{13} & s_{23} & s_{33} & 0 & s_{35} & 0 \\ 0 & 0 & 0 & s_{44} & 0 & s_{46} \\ s_{15} & s_{25} & s_{35} & 0 & s_{55} & 0 \\ 0 & 0 & 0 & s_{46} & 0 & s_{66} \end{vmatrix} \quad \begin{vmatrix} 0 & 0 & 0 & d_{14} & 0 & d_{16} \\ d_{21} & d_{22} & d_{23} & 0 & d_{25} & 0 \\ 0 & 0 & 0 & d_{34} & 0 & d_{36} \end{vmatrix} \quad \begin{vmatrix} \varepsilon_{11} & 0 & \varepsilon_{13} \\ 0 & \varepsilon_{22} & 0 \\ \varepsilon_{13} & 0 & \varepsilon_{33} \end{vmatrix}$$

Class $3m$

($LiNbO_3$, $LiTaO_3$, *Tourmaline*)

$$\begin{vmatrix} s_{11} & s_{12} & s_{13} & s_{14} & 0 & 0 \\ s_{12} & s_{11} & s_{13} & -s_{14} & 0 & 0 \\ s_{13} & s_{13} & s_{33} & 0 & 0 & 0 \\ s_{14} & -s_{14} & 0 & s_{44} & 0 & 0 \\ 0 & 0 & 0 & 0 & s_{44} & 2s_{14} \\ 0 & 0 & 0 & 0 & 2s_{14} & x \end{vmatrix} \quad \begin{vmatrix} 0 & 0 & 0 & 0 & d_{15} & -2d_{22} \\ -d_{22} & d_{22} & 0 & d_{15} & 0 & 0 \\ d_{31} & d_{31} & d_{33} & 0 & 0 & 0 \end{vmatrix} \quad \begin{vmatrix} \varepsilon_{11} & 0 & 0 \\ 0 & \varepsilon_{11} & 0 \\ 0 & 0 & \varepsilon_{33} \end{vmatrix}$$

$$xs_{66} = 2(s_{11} - s_{12})$$

Class 32

(*Quartz*, Te, *and* Se)

$$\begin{vmatrix} \text{Same as for } 3m \end{vmatrix} \quad \begin{vmatrix} d_{11} & -d_{11} & 0 & d_{14} & 0 & 0 \\ 0 & 0 & 0 & 0 & -d_{14} & -2d_{11} \\ 0 & 0 & 0 & 0 & 0 & 0 \end{vmatrix} \quad \begin{vmatrix} \varepsilon_{11} & 0 & 0 \\ 0 & \varepsilon_{11} & 0 \\ 0 & 0 & \varepsilon_{33} \end{vmatrix}$$

Class $mm2$

($LiGaO_2$, SbSI *in ferroelectric range*, $Ba_{.4}Na_{.2}NbO_3$, *orthorhombic* $BaTiO_3$)

$$\begin{vmatrix} s_{11} & s_{12} & s_{13} & 0 & 0 & 0 \\ s_{12} & s_{22} & s_{23} & 0 & 0 & 0 \\ s_{13} & s_{23} & s_{33} & 0 & 0 & 0 \\ 0 & 0 & 0 & s_{44} & 0 & 0 \\ 0 & 0 & 0 & 0 & s_{55} & 0 \\ 0 & 0 & 0 & 0 & 0 & s_{66} \end{vmatrix} \quad \begin{vmatrix} 0 & 0 & 0 & 0 & d_{15} & 0 \\ 0 & 0 & 0 & d_{24} & 0 & 0 \\ d_{31} & d_{32} & d_{33} & 0 & 0 & 0 \end{vmatrix} \quad \begin{vmatrix} \varepsilon_{11} & 0 & 0 \\ 0 & \varepsilon_{22} & 0 \\ 0 & 0 & \varepsilon_{33} \end{vmatrix}$$

Class 222

(*Rochelle salt outside ferroelectric range*)

$$\begin{vmatrix} \text{Same as for } mm2 \end{vmatrix} \quad \begin{vmatrix} 0 & 0 & 0 & d_{14} & 0 & 0 \\ 0 & 0 & 0 & 0 & d_{25} & 0 \\ 0 & 0 & 0 & 0 & 0 & d_{36} \end{vmatrix} \quad \begin{vmatrix} \varepsilon_{11} & 0 & 0 \\ 0 & \varepsilon_{22} & 0 \\ 0 & 0 & \varepsilon_{33} \end{vmatrix}$$

Table I (*continued*)

Class $\bar{4}2m$

(*ADP, KDP*)

$$\begin{vmatrix} s_{11} & s_{12} & s_{13} & 0 & 0 & 0 \\ s_{12} & s_{11} & s_{13} & 0 & 0 & 0 \\ s_{13} & s_{13} & s_{33} & 0 & 0 & 0 \\ 0 & 0 & 0 & s_{44} & 0 & 0 \\ 0 & 0 & 0 & 0 & s_{44} & 0 \\ 0 & 0 & 0 & 0 & 0 & s_{66} \end{vmatrix} \quad \begin{vmatrix} 0 & 0 & 0 & d_{14} & 0 & 0 \\ 0 & 0 & 0 & 0 & d_{14} & 0 \\ 0 & 0 & 0 & 0 & 0 & d_{36} \end{vmatrix} \quad \begin{vmatrix} \varepsilon_{11} & 0 & 0 \\ 0 & \varepsilon_{11} & 0 \\ 0 & 0 & \varepsilon_{33} \end{vmatrix}$$

Class $4mm$

(*Crystal* $BaTiO_3$ *in tetragonal range, tetragonal* Pb(Zr, Ti)O_3)

Same as for $\bar{4}2m$

$$\begin{vmatrix} 0 & 0 & 0 & 0 & d_{15} & 0 \\ 0 & 0 & 0 & d_{24} & 0 & 0 \\ d_{31} & d_{31} & d_{33} & 0 & 0 & 0 \end{vmatrix} \quad \begin{vmatrix} \varepsilon_{11} & 0 & 0 \\ 0 & \varepsilon_{11} & 0 \\ 0 & 0 & \varepsilon_{33} \end{vmatrix}$$

Class $6mm$ and ∞m

(CdS, ZnO, CdSe, AlN, BeO) $6mm$

(*Poled ferroelectric ceramics*) ∞m

$$\begin{vmatrix} s_{11} & s_{12} & s_{13} & 0 & 0 & 0 \\ s_{12} & s_{11} & s_{13} & 0 & 0 & 0 \\ s_{13} & s_{13} & s_{33} & 0 & 0 & 0 \\ 0 & 0 & 0 & s_{44} & 0 & 0 \\ 0 & 0 & 0 & 0 & s_{44} & 0 \\ 0 & 0 & 0 & 0 & 0 & x \end{vmatrix} \quad \begin{vmatrix} 0 & 0 & 0 & 0 & d_{15} & 0 \\ 0 & 0 & 0 & d_{15} & 0 & 0 \\ d_{31} & d_{31} & d_{33} & 0 & 0 & 0 \end{vmatrix} \quad \begin{vmatrix} \varepsilon_{11} & 0 & 0 \\ 0 & \varepsilon_{11} & 0 \\ 0 & 0 & \varepsilon_{33} \end{vmatrix}$$

$$xs_{66} = 2(s_{11} - s_{12})$$

Class $\bar{4}3m$

(ZnS, ZnTe, CdTe, ZnSe, GaAs, $Bi_{12}GeO_{20}$)

$$\begin{vmatrix} s_{11} & s_{12} & s_{12} & 0 & 0 & 0 \\ s_{12} & s_{11} & s_{12} & 0 & 0 & 0 \\ s_{12} & s_{12} & s_{11} & 0 & 0 & 0 \\ 0 & 0 & 0 & s_{44} & 0 & 0 \\ 0 & 0 & 0 & 0 & s_{44} & 0 \\ 0 & 0 & 0 & 0 & 0 & s_{44} \end{vmatrix} \quad \begin{vmatrix} 0 & 0 & 0 & d_{14} & 0 & 0 \\ 0 & 0 & 0 & 0 & d_{14} & 0 \\ 0 & 0 & 0 & 0 & 0 & d_{14} \end{vmatrix} \quad \begin{vmatrix} \varepsilon_{11} & 0 & 0 \\ 0 & \varepsilon_{11} & 0 \\ 0 & 0 & \varepsilon_{11} \end{vmatrix}$$

1. The g and d matrices are identical in form.
2. The e and h matrices are identical in form to those for g and d except for classes $3m$, where $e_{16} = -e_{22}$ and 32, where $e_{26} = -e_{11}$.
3. The c and s matrices are identical in form except for classes $3m$ and 32, where $c_{66} = \frac{1}{2}(c_{11} - c_{12})$ and $c_{56} = c_{14}$, and classes $6mm$ and ∞m, where $c_{66} = \frac{1}{2}(c_{11} - c_{12})$.
4. Lithium iodate ($LiIO_3$) is in class 6. Matrices are identical to those for $6mm$ with the exception of additional piezoelectric constants $d_{14} = -d_{25}$.

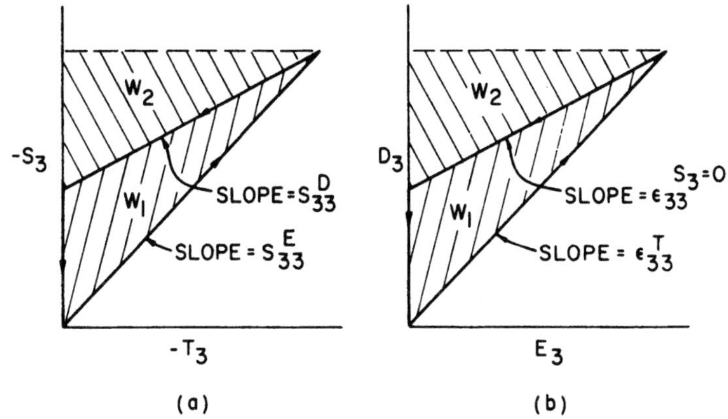

Fig. 1. Illustration of the piezoelectric coupling factor for $k_{33}^l = 0.70$. Here $(k_{33}^l)^2 = W_1/(W_1 + W_2)$.

The slope of the strain–stress curve on compression (negative stress) is s_{33}^E. Before the compressive stress is removed the element is open-circuited and the slope of the strain–stress curve is equal to s_{33}^D as the stress is reduced to zero. The element is connected to an electric load to complete the cycle. As work is delivered to the electric load the internal strain returns to its initial state. For this idealized cycle the work W_1 delivered to the electrical load and the part of the energy W_2 stored on compression and not available to the electrical load in Fig. 1 are related to the coupling factor as follows:

$$(k_{33}^l)^2 = \frac{W_1}{W_1 + W_2} = \frac{s_{33}^E - s_{33}^D}{s_{33}^E} \tag{8}$$

The set of equations (2) can be used to obtain s_{33}^D in terms of the other constants. For this case the set (2) reduces to the following (neglecting purely elastic equations and with $\varepsilon_{mk} = 0$ when $m \neq k$):

$$S_3 = s_{33}^E T_3 + d_{33} E_3 \tag{9}$$

$$D_3 = d_{33} T_3 + \varepsilon_{33}^T E_3 \tag{10}$$

Setting $D_3 = 0$ in Eq. (10), solving for E_3, and substituting in Eq. (9), gives:

$$\left(\frac{S_3}{T_3}\right)_{D=0} = s_{33}^D = s_{33}^E - \frac{d_{33}^2}{\varepsilon_{33}^T} = s_{33}^E\left(1 - \frac{d_{33}^2}{s_{33}^E \varepsilon_{33}^T}\right) \tag{11}$$

Solving Eq. (8) for s_{33}^D,

$$s_{33}^D = [1 - (k_{33}^l)^2]s_{33}^E \tag{12}$$

therefore,

$$(k_{33}^l)^2 = \frac{d_{33}^2}{s_{33}^E \varepsilon_{33}^T} \tag{13}$$

For conversion of energy stored from an electrical source to mechanical work, the cycle shown in Fig. 1b is the reference. Here the specimen is free to cross contract in the plane perpendicular to X_3. It is connected to a dc voltage providing an electric field E_3. Then the piezoelectric element is blocked mechanically as the voltage source is disconnected. The block is then removed and in its place a finite mechanical load is provided. For this idealized cycle the work W_1 delivered to the mechanical load and the part of the energy W_2 stored on application of the voltage source but not available to the mechanical load are related to the coupling factor as follows:

$$(k_{33}^l)^2 = \frac{W_1}{W_1 + W_2} = \frac{\varepsilon_{33}^T - \varepsilon_{33}^{S_3=0}}{\varepsilon_{33}^T} \tag{14}$$

Setting $S_3 = 0$ in Eq. (9), solving for T_3 and substituting in Eq. (10), and then solving Eq. (14) for $\varepsilon_{33}^{S_3=0}$ also gives Eq. (13).

A different coupling factor results if, in Fig. 1, S_1 and S_2 are held at zero (lateral clamping). In this case the material coupling factor k_t results directly using the set of equations (1),

$$(k_{33}^t)^2 = \frac{e_{33}^2}{\varepsilon_{33}^S c_{33}^D} \tag{15}$$

The corresponding cycle is compared in Fig. 2 to that with T_1 and T_2 held at zero, to scale for the commonly used lead titanate zirconate ceramic composition PZT-4. In general $k_{33}^l > k_{33}^t$, but this need not be so.*

A large number of material coupling factors can be defined for different conditions of elastic restraint. Some of the more important material coupling factors for piezoelectric ceramics and crystals of class $6mm$ (e.g., CdS, ZnO) are listed in Table II. Appropriate elastic boundary conditions are listed also. Many of these same material coupling factors exist as well in other crystal classes, but in many cases transformed axes must be chosen. This is discussed in more general terms in Ref. (2).

* Other types of cycles could also be chosen in Figs. 1 and 2, but those described here relate directly to the material coupling factors in general use. See Ref. (13).

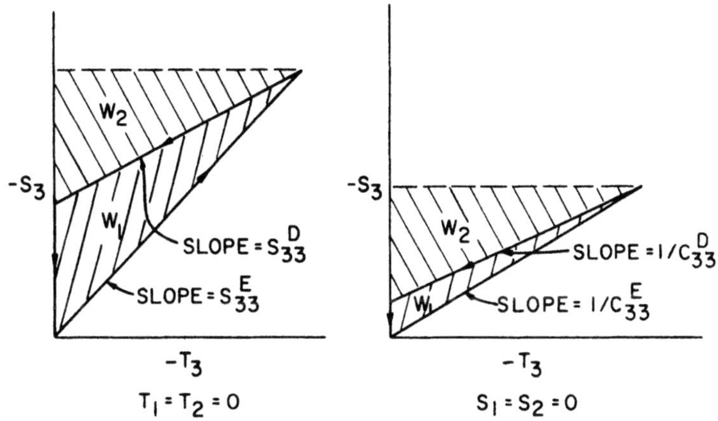

Fig. 2. Illustration of the piezoelectric coupling factors k_{33}^l and k_{33}^t with PZT-4 element free of lateral restraint ($T_1 = T_2 = 0$) and clamped laterally ($S_1 = S_2 = 0$). $k_{33}^l = 0.70$, $k_{33}^t = 0.51$, $s_{33}^E = 15.5 \times 10^{-12}$ m²/N, and $c_{33}^E = 11.5 \times 10^{10}$ N/m².

Table II. Important Material Coupling Factors of Piezoelectric Ceramics or Crystals of Class *6mm* (e.g., CdS and ZnO) and Corresponding Elastic Boundary Conditions

Material coupling factor	Elastic boundary condition	Identification
$k_{31}^l = d_{31}/\sqrt{s_{11}^E \varepsilon_{33}^T}$	All stress components zero except T_1	$l = X_1$, $t = X_3$
$k_{33}^l = d_{33}/\sqrt{s_{33}^E \varepsilon_{33}^T}$	All stress components zero except T_3	$l = X_3$
$k_p = k_{31}\sqrt{2/(1-\sigma_p{}^E)}$	All stress components zero except $T_1 = T_2$	$t = X_3$
$k_{31}^w = (k_{31}/\sqrt{1-k_{31}^2})(\sqrt{(1+\sigma_p{}^E)/(1-\sigma_p{}^E)}$	T_3 and S_2 zero	$w = X_1$, $t = X_3$
$k_{33}^w = (k_{33}+\sigma_m{}^E k_{31})/\sqrt{[1-(\sigma_m{}^E)^2](1-k_{31}^2)}$	T_2 and S_1 zero	$w = X_3$, $l = X_1$
$k_{33}^t = e_{33}/\sqrt{c_{33}^D \varepsilon_{33}^S}$	All strain components zero except S_3	$t = X_3$
$k_{15} = e_{15}/\sqrt{c_{44}^D \varepsilon_{33}^S}$	All strain components zero except S_5	$t = X_1$

1. $\sigma_p{}^E = -s_{12}^E/s_{11}^E$; $\sigma_m{}^E = -s_{13}^E/\sqrt{s_{11}^E s_{33}^E}$; σ_p and σ_m are, respectively, the planar and meridianal Poisson's ratios.
2. k_{31}^l negative; k_{33}^l positive.

3. MODES OF VIBRATION OF PIEZOELECTRIC ELEMENTS

Applications of piezoelectric materials can be divided generally into two classes—resonant and nonresonant. This section is concerned only with the former, and in fact this is the most important category, involving not only resonant transducers but the use of piezoelectric resonators in electric wave filters as well. Nonresonant applications include instrumentation transducers as well as phonograph pickup, microphone, and loudspeaker elements.

Piezoelectric resonators may be classified as to whether they are low-frequency or high-frequency devices. In this section only pure resonant modes will be considered, but it must be recognized that geometrical factors as well as low crystal symmetry can lead to coupling of energy from one mode to another.

3.1. Low-Frequency Modes

Low-frequency modes involve mechanical resonance controlled by a major dimension of the resonating element. Table III lists some of the most common resonant modes which can exist in piezoelectric elements.* The first three can be classified low-frequency modes, since the controlling dimension is the major dimension. The WE_p and WE_t modes may generally also be classified as low-frequency modes, but the controlling dimension is the width rather than the length of the bar. All modes in Table III except PE_t, FS_t, and LE_t-ring may be excited in all piezoelectric crystals, but in many cases rotated cuts are required. The planar mode of a disk can be excited as a pure mode only in piezoelectric ceramics or crystals of classes $6mm$, $3m$, 3, and 6. Mode LE_t-ring with radial E can be excited only in piezoelectric ceramics, but with an axial field it is excited in all crystals in which the planar mode of a disk can be excited.

The face-shear mode (transverse E) is a low-frequency mode with material coupling factor of the form:

$$k_{36}^2 = d_{36}/(\varepsilon_{33}^T s_{66}^E) \tag{16}$$

This mode cannot be excited in piezoelectric ceramics, although the corresponding vibration with E parallel to the surface can. More complex contour modes are not discussed here. There is an extensive literature.[2,14–16]

* The characteristic resonator frequencies f_p and f_s are discussed in Sec. 3.3. These are frequencies of zero admittance and zero impedance respectively for a *lossless* resonator.

Table III. Low-Frequency Resonant Modes

Resonant mode	Electric boundary condition	k_{mat} in terms of f_s and f_p	(Wave velocity)^2
LE_t-bar	Constant E at f_s	$\dfrac{(k_{31}^l)^2}{1-(k_{31}^l)^2} = \dfrac{\pi}{2}\dfrac{f_p}{f_s}\tan\dfrac{\pi}{2}\dfrac{\Delta f}{f_s}$	$1/\varrho s_{11}^E$
LE_t-ring	Constant E at f_s	$(k_{31}^l)^2 = \dfrac{f_p^2 - f_s^2}{f_p^2}$	Symmetric
LE_p	Constant D at f_p	$(k_{33}^l)^2 = \dfrac{\pi}{2}\dfrac{f_s}{f_p}\tan\dfrac{\pi}{2}\dfrac{\Delta f}{f_p}$	$1/\varrho s_{33}^D$
PE_t-disc	Constant E at f_s	See ref. (3) and Fig. 9	$1/\varrho s_{11}^E[1-(\sigma_p{}^E)^2]$
PE_t-sphere	Constant E at f_s	$k_p{}^2 = \dfrac{f_p^2 - f_s^2}{f_p^2}$	Symmetric
WE_t	Constant E at f_s	$\dfrac{(k_{31}^w)^2}{1-(k_{31}^w)^2} = \dfrac{\pi}{2}\dfrac{f_p}{f_s}\tan\dfrac{\pi}{2}\dfrac{\Delta f}{f_s}$	$1/\varrho_{11}^E s[1-(\sigma_p{}^E)^2]$
WE_p	Constant D at f_p	$(k_{33}^w)^2 = \dfrac{\pi}{2}\dfrac{f_s}{f_p}\tan\dfrac{\pi}{2}\dfrac{\Delta f}{f_p}$	$1/\varrho s_{33}^D[1-(\sigma_m{}^D)^2]$
FS_t	Constant E at f_s	$\dfrac{k_{36}^2}{1-k_{36}^2} = \dfrac{\pi}{2}\dfrac{f_p}{f_s}\tan\dfrac{\pi}{2}\dfrac{\Delta f}{f_s}$	$1/\varrho s_{66}^E$

1. The symmetric modes have uniform distribution of stress and are free of overtones.
2. LE_t = longitudinal expander, transverse E; LE_p = longitudinal expander, parallel E; PE_t = planar expander, transverse E; WE_t = width expander, transverse E; WE_p = width expander, parallel E; FS_t = face shear, transverse E; $\sigma_p{}^E = -s_{12}^E/s_{11}^E$; $\sigma_m{}^D = -s_{13}^D/\sqrt{s_{11}^D s_{33}^D}$; ϱ = density. In all cases $k_{eff}^2 = (f_p{}^2 - f_s{}^2)/f_p{}^2$.

Although the analysis of piezoelectric resonators or transducers can be accomplished with the wave equations, the usual method is the equivalent circuit approach first suggested by Mason.[17] The equivalent circuit for each type of piezoelectric transducer can be derived from the wave equation, the appropriate piezoelectric equations, and the mechanical boundary conditions. If the wave propagation is parallel to a major dimension a low-frequency mode exists. If the electroded surfaces are parallel to the propagation direction, Eq. (2) is appropriate (length expander bar with transverse E). If the electroded surfaces are perpendicular to the propagation direction, Eq. (3) is appropriate (length expander bar with parallel E).* The equivalent circuit for the transverse expander bar is shown in Fig. 3. The

* In the general case E will not be parallel to the length. In practical cases ($\beta_{ij} = 0$ when $i \neq j$) it is.

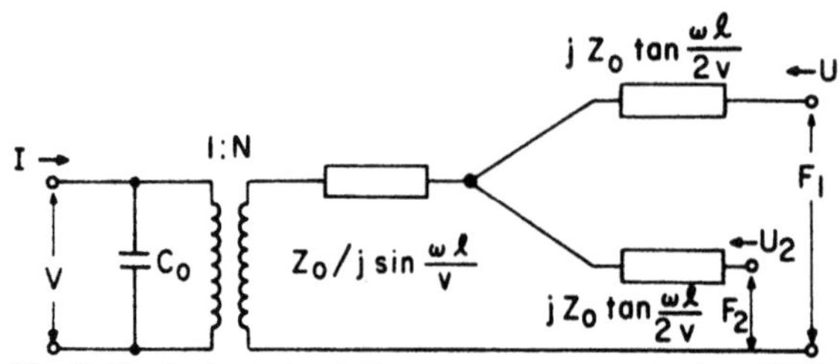

Fig. 3. Equivalent circuit for transverse expander bar (LE_t); $C_0 = lw\varepsilon_{33}^T[1 - (k_{31}^l)^2]/t$, $v = 1/(\varrho s_{11}^E)^{1/2}$, $Z_0 = \varrho wtv$, and $N = wd_{31}/s_{11}^E$.

ideal electromechanical transformer allows the use of electrical and mechanical input or output quantities. The equivalent circuit for the parallel expander bar is shown in Fig. 4. In Figs. 3 and 4, I and V are current and voltage and F and U are force and displacement. Z_0 is the mechanical impedance and N is the transducer ratio (V to F or I to $\dot{U}$). These equivalent circuits are valid representations of the piezoelectric elements from dc to a frequency approaching the fundamental resonance of the width mode of the bar (if excited) including all overtones of the length mode. Accuracy of the equivalent circuits is limited only by the accuracy of the assumed boundary conditions.*

Simplified equivalent circuits which hold in a frequency range near each resonance, i.e., where $\omega l/2v \sim \pi/2$, $3\pi/2$, $5\pi/2$ etc., are obtained by expansion of the impedance functions about the resonant frequency and comparing this with the equivalent impedance functions for series and parallel LC networks. The most useful form of this equivalent representation, that with one vibrating end free (inertia drive) is shown in Fig. 5. The equivalent circuit elements are listed in Table IV. The impedance of

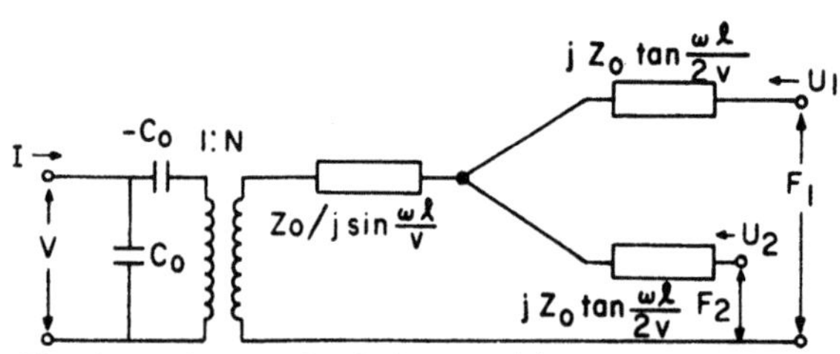

Fig. 4. Equivalent circuit for parallel expander bar (LE_p); $C_0 = wt\varepsilon_{33}^T[1 - (k_{33}^l)^2]/l$, $v = 1/(\varrho s_{33}^D)^{1/2}$, $Z_0 = \varrho wtv$, and $N = wtg_{33}/ls_{33}^E\beta_{33}^T$.

* The different forms of the equivalent circuits Figs. 3 and 4 are due to different boundary conditions. Figure 3 gives mechanical resonance at f_s, while Fig. 4 gives mechanical resonance at f_p.

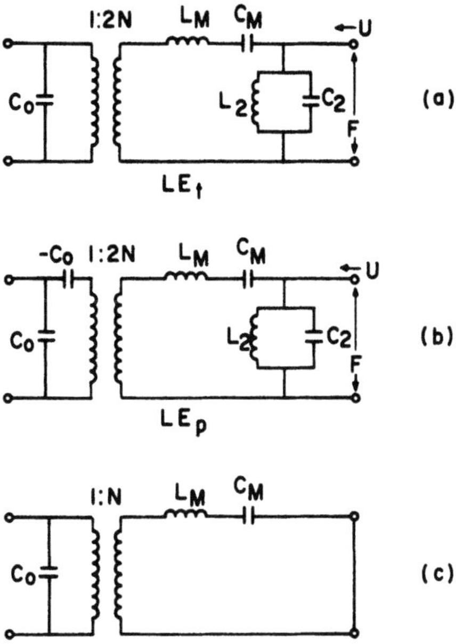

Fig. 5. Equivalent circuits near resonance for a) LE_t mode bar and b) LE_p mode bar each free on one end; equivalent circuit for c) symmetric mode elements. Parameters are listed in Table V.

the shunt circuit L_2C_2 is usually large compared to normally encountered mechanical loads and is generally neglected.

The LE_t-ring and PE_t-sphere modes are unique in that the tangential stress and planar stress, respectively, are uniform; thus, there are no overtones. The equivalent circuit shown in Fig. 5c holds for these modes from dc to a frequency approaching the range of resonance of one of the other dimensions, i.e., the thickness or width of the ring or thickness of the spherical shell. Equivalent circuit parameters are also listed in Table IV.

3.2. High-Frequency or Thickness Modes

Thickness modes exist in thin piezoelectric plates. In pure form there are only thickness-extensional and thickness-shear modes. Both types of modes exist with E perpendicular or parallel to the major surfaces, but modes with E parallel to the plate will not be discussed here. It should be noted, however, that wave velocities are in this case related to the elastic constants at constant E.

Table IV. Equivalent Circuit Parameters for Fig. 5 (Fundamental of Longitudinal Modes)

	Longitudinal Modes	
	LE$_t$	LE$_p$
N	wd_{31}/s_{11}^E	wtd_{33}/ls_{33}^E
ν	$1/\sqrt{\varrho s_{11}^E}$	$1/\sqrt{\varrho s_{33}^D}$
C_0	$\varepsilon_{33}^T(1 - k_{\text{eff}}^2)lw/t$	$\varepsilon_{33}^T(1 - k_{\text{eff}}^2)wt/l$
C_M	$[k_{\text{eff}}^2/(k_{31}^l)^2]s_{11}^E l/4tw$	$[k_{\text{eff}}^2/(k_{33}^l)^2]s_{33}^D l/4tw$
L_M	$4[(k_{31}^l)^2/k_{\text{eff}}^2]twl\varrho/\pi^2$	$4[(k_{33}^l)^2/k_{\text{eff}}^2]twl\varrho/\pi^2$
L_2	$[k_{\text{eff}}^2/(k_{31}^l)^2]twl\varrho$	$[k_{\text{eff}}^2/(k_{33}^l)^2]tlw\varrho$
C_2	$[(k_{31}^l)^2/k_{\text{eff}}^2]ls_{11}^E/\pi^2 wt$	$[(k_{33}^l)^2/k_{\text{eff}}^2]ls_{33}^D/\pi^2 wt$

	Symmetric (Lumped) Modes[a]		
	PE$_t$-spherical shell	LE$_t$-ring, axial E	LE$_t$-ring, radial E
N	$8\pi a d_{31}/s_{11}^E(1 - \sigma^E)$	$2\pi t d_{31}/s_{11}^E$	$2\pi h d_{31}/s_{11}^E$
C_0	$4\pi^2 a^2 \varepsilon_{33}^T(1 - k_p{}^2)/t$	$2\pi a t \varepsilon_{33}^T[1 - (k_{31}^l)^2]/h$	$2\pi h \varepsilon_{33}^T[1 - (k_{31}^l)^2]/t$
C_M	$s_{11}^E(1 - \sigma^E)/8\pi t$	$as_{11}^E/2\pi ht$	$as_{11}^E/2\pi ht$
L_M	$4\pi a^2 t\varrho$	$2\pi a t h\varrho$	$2\pi a t h\varrho$

[a] a = radius, t = radial thickness, and h = axial length of ring.

Table V lists the two types of thickness modes of interest here with some of the pertinent parameters. Although pure thickness modes exist only for plates with infinite lateral dimensions, since otherwise lateral strains are not everywhere zero, these equations apply for practical purposes if lateral dimensions exceed t by an order of magnitude or more. The

Table V. Thickness Modes with E Parallel to Propagation Direction

Resonant mode	Electric boundary conditions	k_{mat} in terms of f_s and f_p	Wave velocity
TE	Constant D at f_p	$(k_{33}^t)^2 = \dfrac{\pi}{2}\dfrac{f_s}{f_p}\tan\dfrac{\Delta f}{f_p}$	$\sqrt{\bar{c}_{33}/\varrho}$
TS	Constant D at f_p	$k_{15}^2 = \dfrac{\pi}{2}\dfrac{f_s}{f_p}\tan\dfrac{\Delta f}{f_p}$	$\sqrt{\bar{c}_{55}/\varrho}$

TE = thickness extensional

TS = thickness shear

$\bar{c}_{33} = c_{33}^E + e_{33}^2/\varepsilon_{33}^S$

$\bar{c}_{55} = c_{55}^E + e_{15}^2/\varepsilon_{11}^S$

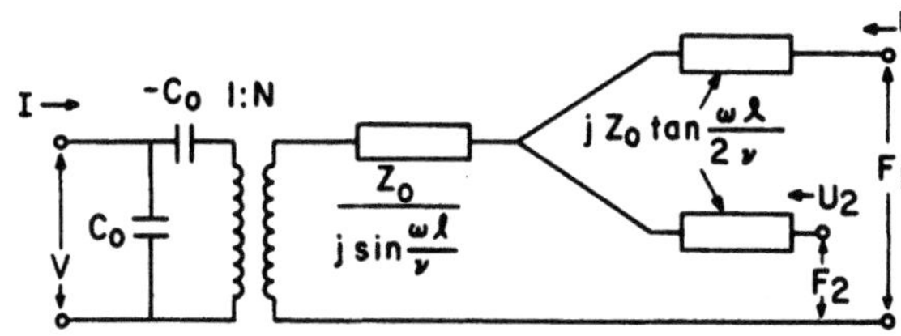

Fig. 6. Equivalent circuit for TE- and TS-modes. Parameters are listed in Table VI.

case where the electrode does not cover the major surfaces of the plate will not be treated here. In this case energy trapping occurs and resonant vibration can proceed independently for isolated resonator areas or with coupling.[18,19] The use of coupled thickness-shear resonators on a single plate in electric wave filters is now well established.[20,21] The equivalent circuit for thickness modes is shown in Fig. 6. The parameters for the two cases at hand are listed in Table VI.

3.3. The Effective Coupling Factor—The Piezoelectric Resonator

The unloaded piezoelectric resonator has an equivalent circuit derived from those in Figs. 3 through 6 simply by short-circuiting the force terminals. The approximate general equivalent circuit which holds in the vicinity of any resonance for the piezoelectric resonator is shown in Fig. 7. Here the resistance R is added on the mechanical side to account for internal mechanical loss; i.e., to account for the imaginary component of the appropriate complex stiffness or compliance coefficient. A resistance to account for the imaginary part of the permittivity is generally not used with the unloaded

Table VI. Equivalent Circuit Parameters of Thickness Modes

	TE	TS
C_0	$A\varepsilon_{33}^S/t$	$A\varepsilon_{11}^S/t$
v	$v_l = \sqrt{\bar{c}_{33}/\varrho}$	$v_s = \sqrt{\bar{c}_{55}/\varrho}$
Z_0	$\varrho A v_l$	$\varrho A v_s$
N	Ae_{33}/t	Ae_{15}/t

A = area of plate

$\bar{c}_{33} = c_{33}^E + e_{33}^2/\varepsilon_{33}^S$

$\bar{c}_{55} = c_{55}^E + e_{15}^2/\varepsilon_{11}^S$

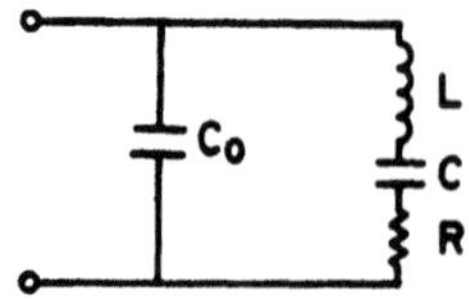

Fig. 7. Piezoelectric resonator equivalent circuit near a given resonance, with $k_{\text{eff}}^2 = (f_p^2 - f_s^2)/f_p^2 = C/(C_0 + C)$ and $Q_M = 1/\omega RC$.

resonator, but this can be important in a well-loaded transducer (see Section 5.2). Since the ideal transformer includes a piezoelectric constant, a rigorous treatment would include the imaginary component of the piezoelectric constant.

The circuit of Fig. 7 has four characteristic frequencies. The frequencies of minimum and maximum impedance respectively are f_m and f_n and those for which the impedance is purely resistive are f_r and f_a. If R were zero only one characteristic set would exist, f_s and f_p, the frequencies of series and parallel resonance. The effective coupling factor k_{eff} is given by

$$k_{\text{eff}}^2 = \frac{f_p^2 - f_s^2}{f_p^2} = \frac{C}{C_0 + C} \tag{17}$$

In the presence of losses f_p and f_s cannot be measured directly, but $\Delta f = f_p - f_s$ can be determined from measured values of $f_a - f_r$ or $f_n - f_m$.[3]

The material coupling factors of Table II are equal to their corresponding effective coupling factors only with the lumped-constant modes, PE_t-spherical shell, and LE_t-ring. With the other modes standing waves exist and elastic and dielectric energies are not coupled uniformly. The effective coupling factor for all resonant modes and the overtones in the absence of intermode coupling is given by Eq. (17). It is related to the material coupling factor through the solution of the boundary value problem for the unloaded resonator. As an example consider Fig. 8, which shows the simplification of the equivalent circuit of Fig. 6 for the fundamental TE mode of a piezoelectric resonator. By writing the equation for Z_{input},

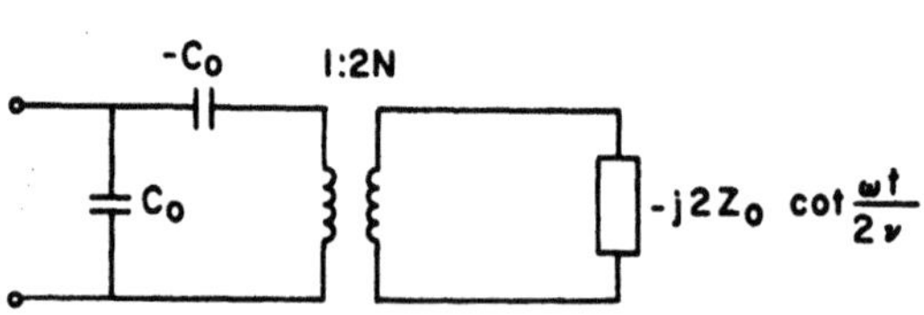

Fig. 8. Simplified equivalent circuit for thickness mode resonators. Parameters are listed in Table VI.

one can readily establish that infinite impedance exists at $f_p = v/2t$. Solution for the zero of impedance at f_s leads to the following equation using the values of N, Z_0 and, v given previously:

$$(k_{33}^t)^2 = \frac{\pi f_s}{2f_p} \tan \frac{\pi \Delta f}{2f_p} \tag{18}$$

For $\Delta f/f_p$ very small the above expression and all others except the lumped-constant modes in Tables III and IV* reduce to the following simplified form,

$$k_{\text{mat}}^2 \sim \frac{\pi^2}{4} \frac{\Delta f}{f_p} \tag{19}$$

and also

$$k_{\text{eff}}^2 \sim 2 \frac{\Delta f}{f_p} \tag{20}$$

Thus for k very small

$$k_{\text{mat}}^2 \sim \frac{\pi^2}{8} k_{\text{eff}}^2 \tag{21}$$

Overtones $f_p(n)$ and $f_s(n)$ of the fundamental mode exist. For modes with elastic wave propagation parallel to the exciting electric field $f_p(n)$ forms an exact odd harmonic sequence, $n = 1, 3, 5, \ldots$. If propagation is perpendicular to the exciting field, $f_s(n)$ forms an exact odd harmonic sequence. *For small values of k* the effective coupling factors of the overtones decrease as $1/n$ and the sum of the squares of the effective coupling factors for fundamental and overtones equals the square of the appropriate material coupling factor.

$$k_{\text{mat}}^2 = k_{\text{eff}}^2\left(1 + \frac{1}{9} + \frac{1}{25} \cdots\right) = \frac{\pi^2}{8} k_{\text{eff}}^2 \tag{22}$$

Figure 9 shows the ratio $k_{\text{mat}}/k_{\text{eff}}$ as a function of k_{eff} for the modes listed in Tables III and V. The appropriate coupling factors may be obtained using Fig. 9 and the measured value of k_{eff}. An alternate method, convenient for TE and TS modes, requires measurement of the fundamental mode f_s and at least one overtone of f_s. The anharmonicity of overtones is readily related to the appropriate material coupling factor.[22,23]

* Strictly speaking the PE_t-disc mode is an exception. In fact, Eq. (18) is a reasonably good expression also in this case.

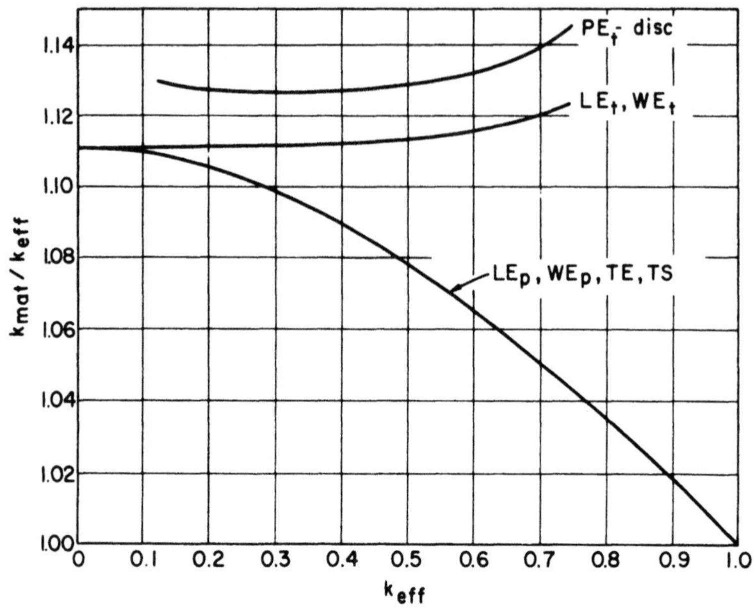

Fig. 9. The ratio k_{mat}/k_{eff} as a function of k_{eff} for important modes of piezoelectric elements.

4. FERROELECTRICITY

4.1. General Description

A ferroelectric crystal is a crystal which has an internal dipole moment, even in the absence of an electric field, which tends to align itself in an allowed direction closest to an applied electric field. The allowed directions are determined by the crystal symmetry. In barium titanate ($BaTiO_3$), for instance, the allowed directions for the electric moment are parallel to an edge, a face diagonal, or a cube diagonal of the reference unit cell in the tetragonal (room temperature), orthorhombic, and rhombohedral states. There are therefore six, twelve, and eight possible orientations, respectively, of the moment in these states.

All ferroelectrics are piezoelectric and pyroelectric. The ferroelectrics are distinguished from other polar piezoelectrics such as tourmaline and cadmium sulfide (CdS) in that the internal moments of the latter cannot be redirected by an external electric field and no nonpolar reference phases are known to exist.

A ferroelectric crystal may, in the unpoled state, have many domains, i.e., small volumes in which the polar moment is of uniform orientation.

The piezoelectric ceramics consist of many multidomain crystallites. When these ceramics are poled domain twinning is drastically reduced, but it is not eliminated. By its very nature ferroelectricity implies nonlinear effects and these effects make piezoelectric ceramics difficult to characterize completely.

4.2. Piezoelectricity in Ferroelectrics—The Piezoelectric Ceramics

Ferroelectrics are the most important class of piezoelectric materials for acoustic applications. They may be classified as those which are piezoelectric only by virtue of the ferroelectric distortion and those which are piezoelectric even in the absence of the ferroelectric distortion. Most useful ferroelectrics, including the piezoelectric ceramics, are in the former class, but Rochelle salt ($NaKC_4H_4O_6 \cdot 4H_2O$) and potassium dihydrogen phosphate (KH_2PO_4) belong to the latter class. An outstanding feature of ferroelectrics is the usual occurrence of very high permittivity and in many cases strong piezoelectric effects. This dielectric enhancement extends well beyond the temperature extremes of the ferroelectric state and accounts for the very strong piezoelectric effects near the Curie points of Rochelle salt and KH_2PO_4. The Curie points of perovskite-structure piezoelectric ceramics constitute an absolute upper temperature limit, since there the ferroelectric distortion disappears, i.e., the structure has a center of symmetry.

The piezoelectric effect of a ferroelectric crystal with a center of symmetry in its reference state above the Curie point may be regarded as electrostriction of the internal field. It has been demonstrated that the piezoelectric effect can be regarded as an electrostrictive phenomenon of the combined applied signal and spontaneous polarization in tetragonal single-crystal barium titanate.[24] In this work the preferred description, a piezoelectric effect, will nevertheless be used.

4.3. Nonlinearities—Domain Effects

Deviations from linear behavior occur in ferroelectrics due to the influence of mechanical and electrical stresses on domain configurations. These interactions may be classified as direct or indirect. Nonlinearities of D–E and S–T are direct and nonlinearities of D–S, E–S, D–T, and E–T are indirect. The direct nonlinearities are more important since they involve dissipation. This is discussed in detail in Section 5. The indirect nonlinearities cause signal distortion. Other important consequences of domain effects

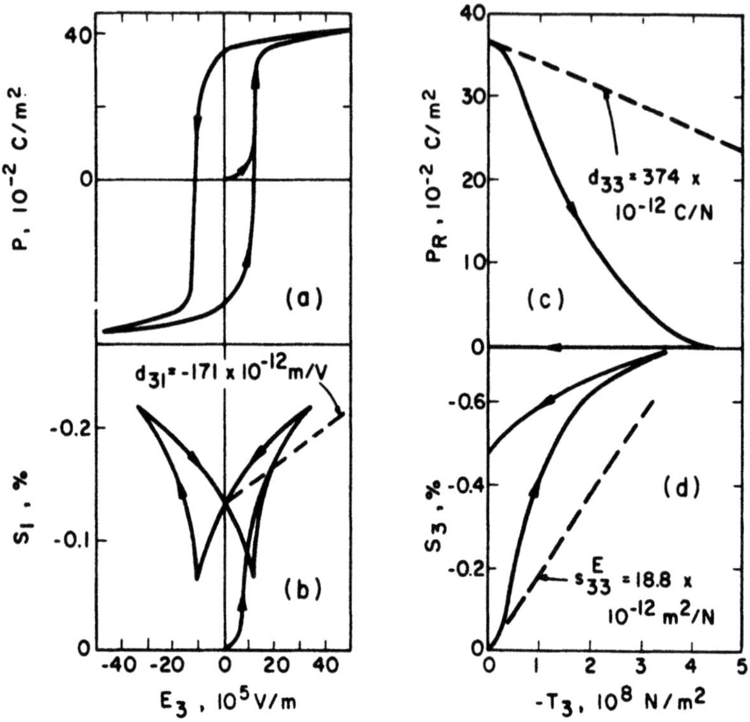

Fig. 10. Nonlinearities in piezoelectric ceramics. a) Polarization (P_3) *vs* electric field (E_3), b) lateral strain (S_1) *vs* electric field (E_3), c) remanent polarization (P_R) *vs* axial compressive stress (T_3), and d) parallel strain (S_3) *vs* axial compressive stress (T_3). All curves for PZT-5A.

are changes of the low signal values of electroelastic constants with high static stress or electric field (Section 7.2.2).

Another domain effect is aging. The domain alignment which occurs during poling leaves the piezoelectric ceramic in a state of high internal stress due to the strain accommodations which must occur. The gradual relief of the internal stress by domain reorientation causes changes in the electroelastic constants. These changes generally occur in roughly equal increments per decade of time and are therefore usually described in these terms. A renewed aging cycle occurs as a result of a major mechanical or electrical disturbance, a factor which complicates characterization further (Section 7.2.1).

Figure 10 illustrates both direct and indirect nonlinearities for PZT-5A*

* PZT-5A is one of many modified lead titanate zirconate compositions. It is described in Section 7.1.

for the extreme case of poling, i.e., for an electric field excursion in excess of the coercive field with a virgin element and for depolarization from this poled state by high-axial compression ($D_3 - T_3$) and ($S_3 - T_3$). The piezoelectric ceramics differ considerably with respect to thresholds for the onset of nonlinearities. This is discussed in detail in Sections 5 and 7.

The inherent nonlinearities of the piezoelectric ceramics and ferroelectric crystals lead to difficulty in characterization. With very low electrical or mechanical drive the piezoelectric ceramics may be considered linear, even though reversible domain-wall motion enhances the piezoelectric and dielectric constants and increases compliance. With increasing electrical or mechanical drive there is a disproportionate increase in piezoelectric response. High electric drive also results in a more than linear increase in D and an increase in dielectric loss. High mechanical drive similarly produces a disproportionate increase in mechanical displacement and an increase in mechanical loss. These effects are frequency dependent, being most severe under very low frequency or static conditions.

4.4. Phase Transitions

Ferroelectrics, with the exception of a few which are subject to decomposition within their ferroelectric range, exhibit phase transitions. The most important transitions are those between ferroelectric and nonpolar states. With all materials which have a centrosymmetric nonpolar state the transition to the nonpolar state marks the disappearance of the piezoelectric effect. Ferroelectrics are also subject to transitions in which the crystallographic orientation of the polar axis changes discontinuously. These transitions are almost always marked in piezoelectric ceramics by peaks in piezoelectric coupling, compliance, and permittivity, and in dielectric and mechanical loss.

The phase diagram for the lead titanate zirconate [$Pb(Zr, Ti)O_3$] system is shown in Fig. 11. Three ferroelectric and two antiferroelectric states are shown as well as the reference paraelectric state. The antiferroelectric states have multiple distorted unit cells in which one can consider the distortions exactly canceled so that there is no net polar moment.

Phase transitions can occur not only with temperature change, as in Fig. 11, but also with mechanical stress and electric field.[25] This will not be discussed here in detail. An electric field favors the ferroelectric state over the nonpolar states. In general, hydrostatic or unidirectional compression favor antiferroelectricity and unidirectional tension favors ferroelectricity. It is possible to utilize these transitions for nonlinear energy

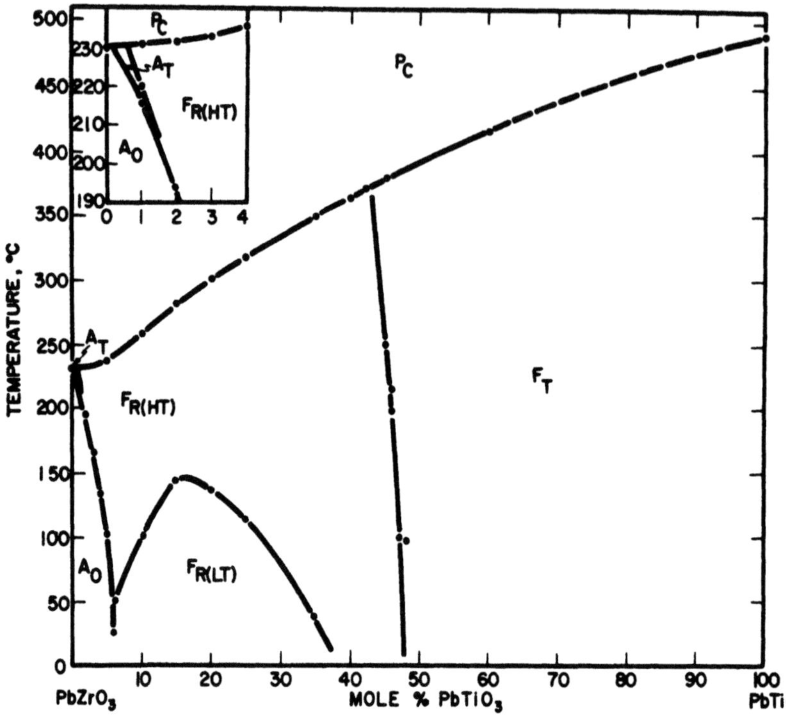

Fig. 11. Lead zirconate–lead titanate phase diagram. Insert is magnified portion of main field. F_T is ferroelectric with tetragonal symmetry; $F_R(LT)$ and $F_R(HT)$ are ferroelectric both with apparent rhombohedral symmetry (not yet distinguished); A_O and A_T are antiferroelectric with orthorhombic and tetragonal symmetry; P_C is paraelectric with cubic symmetry.

conversion, but this is not practical at high frequencies due to hysteresis-type losses.[13,26]

5. DISSIPATION IN PIEZOELECTRIC MATERIALS

5.1. General

Mechanical and dielectric losses exist in all piezoelectric materials. The respective loss factors can be considered to represent the ratio of the imaginary to the real part of the complex dielectric or elastic constant.

$$\tan \delta_M = \frac{s''}{s'} \text{ or } \frac{c''}{c'} \tag{23}$$

$$\tan \delta_E = \frac{\varepsilon''}{\varepsilon'} \text{ or } \frac{\beta''}{\beta'} \tag{24}$$

The appropriate elastic or dielectric constant depends on the elastic and electric boundary conditions (Tables III and V). To account completely for losses which can exist the appropriate piezoelectric loss tangent must also be considered.

$$\tan \delta_{ME} = \frac{d''}{d'}, \frac{e''}{e'}, \frac{g''}{g'}, \text{ or } \frac{h''}{h'} \tag{25}$$

With piezoelectric ceramics there is the additional complication that all loss factors are amplitude dependent, $\tan \delta_E$ on electric field, $\tan \delta_M$ on mechanical stress and $\tan \delta_{ME}$ on both mechanical stress and electric field. Only dielectric and mechanical loss factors in the piezoelectric ceramics will be considered here since there is no information available on high-amplitude piezoelectric loss factors.[27–29]

The variation of dielectric loss factor $(\varepsilon_{33}^T)''/(\varepsilon_{33}^T)'$ with electric field is determined at low frequencies and thus there is no significant internal mechanical stress. The variation of mechanical loss factor with mechanical stress is usually determined from the measured mechanical quality factor Q of an LE_t-ring in resonant vibration, and thus the electric field is very low. One is also generally concerned with the changes in the real values of the elastic and dielectric constants as well.

In this section these quantities are given as functions of ac amplitude with no bias electric field or stress. Mechanical parameters are shown as functions of a one-dimensional stress T_1 at 50–100 kHz and dielectric parameters are shown as functions of the electric field E_3 generally at 1 kHz. Since these losses are generally of the hysteresis type, variations with frequency are not severe at frequencies outside the region of major relaxations

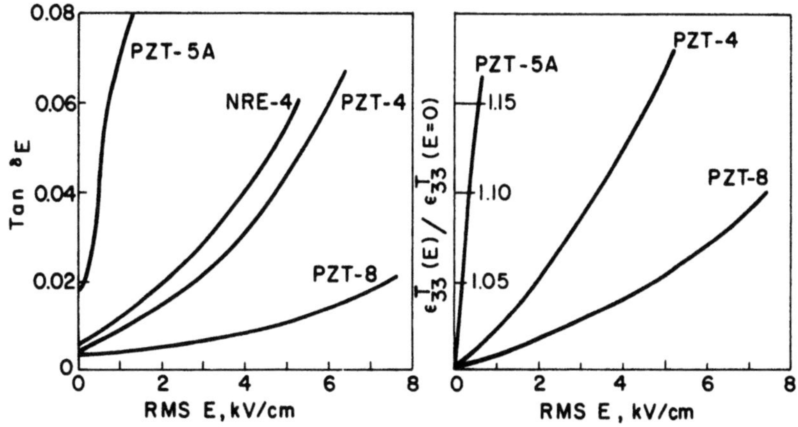

Fig. 12. Tan δ_E and ε_{33}^T *vs* electric field for several piezoelectric ceramics.

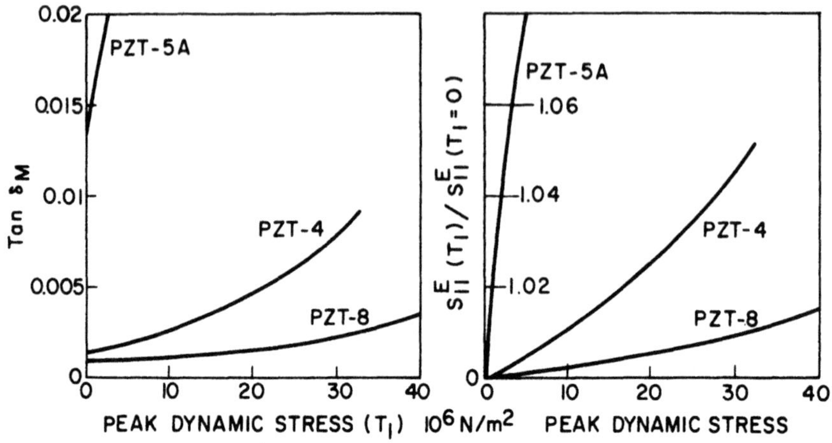

Fig. 13. Tan δ_M and s_{11}^E *vs* dynamic stress (T_1) for several piezoelectric ceramics.

(about 100 MHz). In general, however, loss factors at high amplitudes decrease somewhat with increasing frequency due to domain inertia effects.

Figure 12 shows tan δ_E and ε_{33}^T as functions of electric field for several technically important ceramic materials. These materials are discussed further in Section 7.1. Figure 13 shows tan δ_M and relative changes in s_{11}^E as functions of dynamic stress for some of the same materials. The wide variety in behavior for the different ceramics is immediately evident. This is discussed further in Section 7.

5.2. Effects on Transducer Efficiency and Power Capacity

The effect of internal losses on transducer efficiency is severe only with very lightly loaded (high-Q) transducers or transducers operated at high amplitudes. Other losses in the transducer structure may actually exceed those in the piezoelectric material. Even more important than transducer efficiency, however, are power limitations imposed by heat rise due to energy losses in the piezoelectric elements or other mechanical losses in the transducer structure. The lower the duty cycle of the transducer, the higher the instantaneous dissipated power which can be tolerated.

In this section power limitations imposed by the properties of piezoelectric ceramics will be discussed on a comparative basis only. Two types of transducers will be considered, those with relatively low and relatively high Q (say under 10 and over 30, respectively). It will be convenient to consider a type of composite transducer which is in general use in sonar systems. Sonar transducers are of the low-Q type, but the principles involved in both types are readily demonstrated with this structure. The transducer consists of a hollow piezoelectric ceramic structure with end masses and usually a bias rod through the center. The end masses act to lower the resonance frequency of the structure and to provide better acoustic matching between the ceramic and its fluid load. The rod provides a bias compressive stress to prevent fracture of the ceramic element in tension as it is driven at high dynamic stress. The ceramic element is either a hollow radially-poled tube (LE_t) or a series of parallel connected axially poled rings (LE_p).

The solution is usually handled as a one-dimensional one with plane sound waves; this is valid if the transducer is part of a large array. For purposes of this discussion the metal mass loads can be considered infinitely rigid and the mass of the ceramic is neglected. With these assumptions the stress in the ceramic is uniform and $k_{\text{eff}} = k_{\text{mat}}$ (k_{31}^l or k_{33}^l for the two configurations considered).

The equivalent circuit for equal end masses M and equal fluid loads at each mass is shown in Fig. 14. Actual transducers usually are fluid loaded on only one side, and the end masses can be adjusted independently to change the transducer Q. The simplified arrangement of the circuit Fig. 14, however, is more readily analyzed to point out the important characteristics of the piezoelectric ceramic material. Table VII lists the

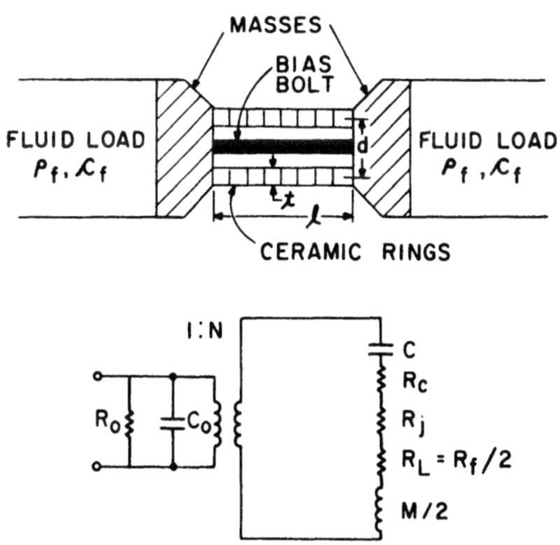

Fig. 14. Equivalent circuit for symmetric sonar-type transducer. The stiffness and compliance of the bias bolt, the compliance of the end masses, and the mass of the ceramic elements are neglected in the simplified equivalent circuit. Parameters are listed in Table VII.

Table VII. Equivalent Circuit Parameters for Sonar-type Composite Transducer

$$R_f = \varrho_f c_f A_r$$

LE_p	LE_t
$C_0 = n^2 A_c \varepsilon_{33}^T [1 - (k_{33}^l)^2]/l$	$C_0 = \varepsilon_{33}^T [1 - (k_{31}^l)^2] \pi dl/t$
$C = s_{33}^E l/A_c$	$C = s_{11}^E l/A_c$
$N = d_{33} A_c n/ls_{33}^E$	$N = d_{31} A_c/ts_{11}^E$

A_c = cross-sectional area of ceramic
A_r = cross-sectional area of end masses
n = number of ceramic rings ($n = 1$ for LE_t)
l = length of rings (total)
t = wall thickness of rings
d = average diameter of rings
ϱ_f, c_f = density and wave velocity of fluid load.

parameters of the equivalent circuit. The resistances R_L, R_c, and R_j in Fig. 14 represent respectively the acoustic load (mass loading neglected), internal mechanical losses in the ceramic, and other mechanical losses. The resistance R_0 represents internal dielectric losses.

$$R_c = \frac{1}{\omega_r C} \tan \delta_M \tag{26}$$

$$R_0 = \frac{(1 - k^2)}{\omega C_0 \tan \delta_E} \tag{27}$$

The term $(1 - k^2)$ in the expression for R_0 occurs because $\tan \delta_E$ is always measured at frequencies well below resonance and is therefore related to the free dielectric constant. Note that R_c and R_0 are actually dependent upon the input voltage and the stress developed in the ceramic respectively. The transducer ratio N gives the force equivalent of voltage. It is modified from its value in Table IV by the factor n because the effective electric field is n times as great.

The dissipated power P_D per unit volume of the ceramic is given by:

$$P_D = P_{DE} + P_{DM} = \omega_r (E_3{}^2 \varepsilon_{33}^T \tan \delta_E + T^2 s^E \tan \delta_M) \tag{28}$$

where P_{DE} and P_{DM} represent internal dielectrically and mechanically dissipated power respectively and E_3 and T are rms values (T_3 and s_{33}^E

are used for LE_p and T_1 and s_{11}^E for LE_t). A loaded Q for the transducer is defined as

$$Q_L = 1/\omega C(R_L + R_j + R_c) \qquad (29)$$

The stress T is then given in terms of E_3 as follows:

$$T = E_3 Q_L\, d/s^E \qquad (30)$$

where appropriate subscripts are used (d_{31}, s_{11}^E or d_{33}, s_{33}^E). Therefore

$$P_{DM} = \omega E_3{}^2 Q_L{}^2\, d^2 \tan \delta_M/s^E = \omega E_3{}^2 Q_L{}^2 k^2 \varepsilon_{33}^T \tan \delta_M \qquad (31)$$

$$P_{DE} = \omega E_3{}^2 \varepsilon_{33}^T \tan \delta_E \qquad (32)$$

The power P per unit volume delivered to the mechanical and acoustic elements (R_L, R_j, and R_c) is given by

$$P = \omega E_3{}^2 k^2 \varepsilon_{33}^T Q_L \qquad (33)$$

Neglecting changes of tan δ_M and tan δ_E with amplitude, therefore, P_{DM}/P is proportional to Q_L while P_{DE}/P is inversely proportional to Q_L for a given value of k. It is, therefore, clear that internal mechanical losses are more important in transducers with high Q_L, dielectric losses with low Q_L.

As an example, suppose that P is set at 25 W/cm³ at 10 kHz. Suppose further that $Q_L = 3$ in one case and 30 in another, corresponding roughly to sonar and ultrasonic cleaning transducers, respectively. Table VIII shows values of electric field and internal stress required for PZT-8, PZT-4, and PZT-5A transducers (Fig. 14). The advantages of the PZT-8 material and the LE_p mode are clear.

6. PARAMETERS OF IMPORTANT PIEZOELECTRIC CRYSTALS

Piezoelectric crystals may be classified in several different ways, for instance according to their use at high or low frequencies, whether they are ferroelectric or nonferroelectric, or whether their properties have been known for some time or only recently discovered. It turns out that the historical classification is most convenient and is therefore used here.

6.1. Older Piezoelectric Crystals

Table IX summarizes the most important properties of common cuts of older piezoelectric crystals. Complete sets of electroelastic coefficients

Table VIII. Values of Driving Electric Field, Stress, and Dissipated Power for Transducer of Fig. 14 with $Q_L = 3$ and 30 and $P = 25$ W/cm³ at 10 kHz

Material	Mode	E_3(rms), kV/cm		T_3(rms), 10^6 N/m²		tan δ_M		tan δ_E	
		$Q_L = 3$	$Q_L = 30$	$Q_L = 3$	$Q_L = 30$	$Q_L = 3$	$Q_L = 30$	$Q_L = 3$	$Q_L = 30$
PZT-8	LE_p	1.9	0.61	9.5	30	0.0012	0.004	0.005	0.004
PZT-8	LE_t	4.1	1.31	10.3	32.5	0.0012	0.004	0.009	0.005
PZT-4	LE_p	1.55	0.49	8.7	27.5	0.003	0.013	0.012	0.006
PZT-4	LE_t	3.3	1.05	9.9	31	0.0032	0.016	0.025	0.009
PZT-5A	LE_p	1.34	0.42	8.1	—[a]	0.05	—[a]	0.08	0.035
PZT-5A	LE_t	2.7	0.87	8.7	—[a]	0.05	—[a]	>0.2	0.07

Material	Mode	P_{DM}, W/cm³		P_{DE}, W/cm³		Total input power $P + P_{DE}$, W/cm³		Total output power with $R_j = 0$ $P - P_{DM}$, W/cm³		Efficiency, %	
		$Q_L = 3$	$Q_L = 30$	$Q_L = 3$	$Q_L = 30$	$Q_L = 3$	$Q_L = 30$	$Q_L = 3$	$Q_L = 30$	$Q_L = 3$	$Q_L = 30$
PZT-8	LE_p	0.09	3.0	0.10	0.01	25.10	25.01	24.91	22.0	99.2	88.0
PZT-8	LE_t	0.09	3.7	0.83	0.037	25.83	25.04	24.91	21.3	96.3	85.2
PZT-4	LE_p	0.23	9.9	0.20	0.01	25.20	25.01	24.77	15.1	98.3	60.3
PZT-4	LE_t	0.24	12.0	2.08	0.068	27.1	25.07	24.76	13.0	91.3	51.8
PZT-5A	LE_p	3.75	—[a]	1.35	—[a]	26.35	—[a]	21.25	—[a]	80.8	—[a]
PZT-5A	LE_t	3.75	—[a]	>14.5	—[a]	>39.5	—[a]	21.25	—[a]	<54	—[a]

Note: Output acoustic power per unit volume of ceramic is $P - P_{DM}$; total input power per unit volume is $P + P_{DE}$ (R_j assumed zero).
[a] For PZT-5A, high amplitude Q_M drops below 30.

Table IX. Properties of Common Cuts of Older Piezoelectric Crystals

IRE notation[1]	Common notation	Mode	k	d, pC/N	e, C/m^2	$\varepsilon/\varepsilon_0$	v, m/s	T.C., $10^{-6}/°C$	Density, $10^3\ kg/m^3$
Quartz									
x	X	TE	0.10	2.3	0.17	4.6	5700	−20	2.65
xy	Y-bar	LE_t	−0.10	−2.3	−0.17	4.6	5400	−2 to −5	2.65
y	Y	TS	−0.14	−4.6	−0.17	4.6	3850	+90	2.65
(yzw) + 35°15′	AT	TS	−0.088	−3.4	−0.095	4.6	3320	0	2.65
(yzw) + 31°	AC	TS	−0.10	−3.7	−0.11	4.6	3300	+20	2.65
(yzw) − 59°	BC	TS	−0.04	−0.9	−0.064	4.7	5000	−25	2.65
$NH_4H_2PO_4$(*ADP*)						$\varepsilon_{33}^T/\varepsilon_0$			
(zxt)45°	Z-45	LE_t	0.28	24	0.30	15.3	3250	−300	1.80
$Li_2SO_4 \cdot H_2O$						$\varepsilon_{22}^S/\varepsilon_0$			
y	Y	TE	0.30	16.3	0.66	9	5470	−300	2.06
Rochelle Salt, 30°C						$\varepsilon_{11}^T/\varepsilon_0$			
(xyt)45°	X-45	LE_t	0.65	275	3.0	350	3100	$+3 \times 10^4$	1.77
						$\varepsilon_{22}^T/\varepsilon_0$			
(yzt)45°	Y-45	LE_t	0.32	30	0.16	9.4	2340	−300	1.77

Note: The symbol "p" in units for d is pico ($= 10^{-12}$).

are listed in Ref. (8) and the data in Table IX related to rotated axes are obtained using the equations listed by Cady.[4]

The most important crystal in Table IX is quartz. Its major application is in frequency control and electric wave filters, and in this field there are no serious contenders. Its unique position results from the extraordinary coincidence of zero first- and second-order temperature coefficients at room temperature for the AT-cut. The AT-cut also has only very weak cross coupling and a piezoelectric coupling factor that is not drastically reduced from the maximum value for quartz. Finally, quartz has a mechanical Q that is high in comparison with other piezoelectric crystals. Synthetic quartz crystals became commercially available about a decade ago, and these are widely used in electric wave filters.

As an ultrasonic transducer or as a delay medium in the range about 10 MHz to 1 GHz, quartz will probably be largely displaced by materials of higher piezoelectric coupling or deposited transducers on the one hand and higher Q media on the other. The availability of large high-quality quartz crystals at low cost will, however, assure continued use.

Properties of the most important quartz cuts are listed in the table. All but the LE_t mode of the Y-bar are high-frequency modes. Of the orientations listed only the AT-cut is used for frequency control, but there are several other orientations with high stability in thickness-shear or low-frequency flexure.[30] The TS mode in Y-cut quartz has the highest piezoelectric coupling, but the elastic cross coupling present in trigonal crystals causes the direction of sound propagation to diverge from the wave normal. When quartz is both the transducer and propagation medium for shear waves the AC- or BC-cuts, for which cross coupling is zero, are chosen.

$NH_4H_2PO_4$(ADP) and $NaKC_4H_4O_6 \cdot 4H_2O$ (Rochelle salt) crystals are commonly used in longitudinal modes. Rochelle salt is also used in flexural modes in phonograph pickups. Rochelle salt is ferroelectric along the X-axis between its -18 and $+23$°C Curie points. Ferroelectricity enhances the piezoelectric effect for the X-cut plate well beyond this range and accounts for the severe temperature dependence of s_{44}^E, ε_{11}^T and d_{14}.* Y-plates are often used when better stability is required.

Lithium sulfate shares with ADP and Rochelle salt an acoustic impedance much better matched to water or organic liquids than either quartz or the piezoelectric ceramics, but it is most often used in the high-

* The compliance s_{44}^D and piezoelectric constant g_{14} do not exhibit significant changes related to the ferroelectricity.

frequency TE-mode, especially as a transducer for ultrasonic flaw detection. It has a uniquely high response to hydrostatic pressure ($d_{22} + d_{21} + d_{23} = 13.5 \times 10^{-12}$ C/N).

6.2. Newer Piezoelectric Crystals

During about the last decade a large number of nonwater-soluble piezoelectric crystals have been grown and their electroelastic constants determined. During the early part of this period, and in fact for a good many years previously, a major goal of this work was the discovery of an alternative to quartz, with higher piezoelectric coupling, for use in electric wave filters. In some cases, the goal was also an alternative to silicon as a high-temperature semiconductor. In neither case were the goals achieved, but several crystals with piezoelectric characteristics were found. During about the last five years the search for crystals with interesting electrooptic characteristics has led to a further proliferation of piezoelectric crystals. This search was generally aimed at crystals of high permittivity, since it is these which have the lowest half-wave retardation voltages,[31] and has yielded the interesting ferroelectrics lithium niobate[32] ($LiNbO_3$), lithium tantalate[33] ($LiTaO_3$), and barium sodium niobate[34] ($Ba_{.4}Na_{.2}NbO_3$). Many of the crystals discussed here meet the need for relatively low permittivity with higher piezoelectric coupling than quartz and favorable mechanical properties.

A number of the compound Group II–Group VI semiconductors crystallize in the cubic-sphalerite and hexagonal-wurtzite structures. With the exception of wurtzite-structure aluminum nitride (AlN), the III–V semiconductors crystallize in the sphalerite structure and have weaker piezoelectricity than the wurtzite structure II–VI compounds. Deposited films of the wurtzite structure II–VI compounds are used as delay-line transducers[35,36] and drivers for composite resonators.[36a] Depletion layer[37] and diffusion layer[38] transducers and acoustic amplification[39] have been demonstrated with cadmium sulfide and cadmium selenide, but these devices now seem to be of little practical value.

Table X lists important parameters of the wurtzite structure compounds (6*mm* symmetry) and crystals of 4*mm*, *mm*2, and 6 symmetry. Lithium gallate ($LiGaO_2$) is properly discussed in connection with the wurtzite structure materials, since the structure[47] consists of alternating trivalent gallium and monovalent lithium ions tetrahedrally surrounded by oxygen. Barium titanate and barium sodium niobate are ferroelectric with Curie points at 120 and 560°C, respectively. Lithium iodate ($LiIO_3$)

Table X. Piezoelectric Properties of Crystals in Symmetry Classes 6*mm*, 4*mm*, *mm*2

Crystal	Crystal class	k_{33}^{l}	k_{31}^{l}	k_{15}	k_{33}^{t}	$\varepsilon_{33}^{T}/\varepsilon_0$	$\varepsilon_{33}^{S}/\varepsilon_0$	$\varepsilon_{11}^{T}/\varepsilon_0$
BeO(40)	6*mm*	0.019	−0.009	—	—	7.66	7.66	—
ZnO(41,42,9)a	6*mm*	0.41	−0.19	−0.31	0.28	11.0	8.84	9.26
		0.480	−0.182	−0.196	0.38	10.9	8.45	8.50
CdS(43)	6*mm*	0.262	−0.119	−0.188	0.154	10.3	9.53	9.35
CdSe(43)	6*mm*	0.194	−0.084	−0.130	0.124	10.6	10.2	9.70
AlN(44)	6*mm*	∼0.30	—	—	∼0.20	9.3	8.5	—
$BaTiO_3$(24)	4*mm*	0.56	−0.315	0.57	—	168	110	2920
$Ba_{.4}Na_{.2}NbO_3$(45)b	*mm*2	—	—	0.21	0.57	—	30	232
				0.25				242
$LiGaO_2$(46,9)b	*mm*2	∼0.28	−0.12	−0.17	0.25	9.2	8.5	7.2
			−0.14	−0.17				6.2
$LiIO_3$(81,82)	6	—	—	0.60	0.51	—	6	12.5

Table X (*continued*)

Crystal	Crystal class	$\varepsilon_{11}^{S}/\varepsilon_0$	d_{33}	d_{31}	d_{15}	e_{33}	e_{31}	e_{15}
				pC/N			C/m²	
BeO[40]	6*mm*	—	0.24	−0.12	—	—	—	—
ZnO[41,42,9]a	6*mm*	8.33	10.6	−5.2	−13.9	1.14	−0.61	−0.59
		8.16	12.4	−5.0	−8.3	1.57	−0.36	−0.36
CdS[43]	6*mm*	9.02	10.3	−5.2	−14.0	0.44	−0.24	−0.21
CdSe[43]	6*mm*	9.53	7.8	−3.9	−10.5	0.35	−0.16	−0.14
AlN[44]	6*mm*	—	5	—	—	—	—	—
$BaTiO_3$[24]	4*mm*	1970	85.6	−34.5	392	—	—	21
$Ba_{.4}Na_{.2}NbO_3$[45]b	*mm*2	222	—	—	—	4.15	—	2.4
		227						3.0
$LiGaO_2$[46,9]b	*mm*2	7.0	7.6	−2.9	−6.1	0.88	—	−0.29
		6.0		−3.8	−5.0			−0.29
LiO_3[81,82]	6	8	—	—	—	1.03	—	0.85

Table X (*continued*)

Crystal	s_{33}^E	s_{11}^E	c_{33}^E	c_{11}^E	c_{44}^E	c_{66}	c_{33}^D	c_{11}^D	c_{44}^D	v_{TE}	v_{TS}	Density,
	pm²/N		10^{10} N/m²							m/sec		10^3 kg/m³
BeO	—	2.30	—	—	—	—	—	—	—	>12000	—	3.01
ZnO	6.94	7.86	21.1	21.0	4.25	4.43	22.9	21.5	4.72	6400	2945	5.68
CdS	17.0	20.7	9.38	9.07	1.50	1.63	9.62	9.13	1.56	4500	1800	4.82
CdSe	17.3	23.4	8.36	7.40	1.32	1.44	8.48	7.42	1.34	3860	1540	5.68
AlN	~2.8	—	~34	—	—	—	~35	—	—	10400	—	3.26
$BaTiO_3$	15.7	8.05	—	—	5.43	11.3	—	—	8.06	5440	3660	6.01
$Ba_{.4}Na_{.2}NbO_3$	—	—	13.5	—	6.68	—	20.0	—	7.00	6150	3640	5.30
					6.65				7.10		3660	
$LiGaO_2$	—	7.2	15.5	—	4.71	—	16.3	—	4.83	6260	3390	4.19
		9.1			5.55				5.70		3690	
$LiIO_3$	—	—	5.67	—	1.82	—	7.67	—	2.84	4130	2520	4.47

[a] Piezoelectric and dielectric data for ZnO – top line Ref. (9), bottom line Ref. (41).
[b] For $LiGaO_2$ and $Ba_{.4}Na_{.2}NbO_3$: k_{31}, k_{15}, ε_{11}, d_{31}, d_{15}, e_{31}, e_{15}, c_{55}, s_{11} listed first and k_{32}, etc., second.

is discussed thoroughly in the chapter by A. H. Meitzler. Unlike all other crystals in Table X it is grown from water solution. It belongs to point group 6 and has an electroelastic matrix set identical to that for class 6*mm* except for two additional piezoelectric constants $d_{14} = -d_{25}$.

Table XI lists data for sphalerite-structure II–VI and III–V compounds as well as bismuth germanate ($Bi_{12}GeO_{20}$).

There is a regular progression of piezoelectric effects in the II–VI compounds of zinc and cadmium with oxygen, sulfur, selenium, and tellurium.[43] The strength of the effect increases with increasing weight of the group II element and decreasing weight of the group VI element. An elementary atomic model assuming constant bond length during elastic deformation of cubic-sphalerite and hexagonal-wurtzite structures leads to a simple relationship between the piezoelectric *e* constants and the effective atomic charge. The effective charge increases with increasing weight of the cation and decreasing weight of the anion.

$LiNbO_3$, $LiTaO_3$, and $Ba_{.4}Na_{.2}NbO_3$ are ferroelectric crystals with very strong piezoelectric effects and favorable mechanical characteristics. Table XII lists pertinent parameters for $LiNbO_3$ and $LiTaO_3$. They are in crystal class 3*m* and, therefore, do not in general sustain pure shear or longitudinal vibrations. Pure thickness extensional modes exist, however, for Z-plates. The strong thickness shears in X- and Y-plates are not pure. This is discussed in more detail in Ref. (52). $Ba_{.4}Na_{.2}NbO_3$ (Table X) is orthorhombic *mm*2 and thus X- and Y-plates have pure thickness shear modes and Z-plates have a pure thickness extensional mode. Lithium tantalate X-plates have a zero first-order frequency–temperature coefficient at -10°C,[53] but this is not likely to become technologically important, since the overtone modes have the usual negative temperature coefficient (~ -40 ppm/°C).[54] It is likely that the anomalous behavior at the fundamental is linked in some way to the impurity of the thickness quasi-shear mode. Lithium niobate has a very high Curie point ($\sim$1200°C) and, therefore, has constants which vary little with temperature over a wide range. It holds promise for use in accelerometers at very high temperature, particularly with charge amplifiers which minimize effects of conductivity.

The ferroelectric semiconductor antimony sulfur iodide (SbSI)[55,56] has a very strong piezoelectric effect below its 22°C Curie point.[57] SbSI generally is grown as polycrystalline aggregates with polar axes aligned but with random orientation about the *Z*-axis, and thus the crystal symmetry of *mm*2 is replaced by ∞m characteristic of ceramics. For the ceramic $k_{33}^t = 0.87$ and $e_{33} = 13$ C/m² at 0°C. Other data at 0°C are $d_h \approx 10^{-9}$ C/N, $\varepsilon_{33}^T/\varepsilon_0 = 2200$, $\varepsilon_{33}^S/\varepsilon_0 = 550$, $c_{33}^E = 1.0 \times 10^{10}$ N/m², and $k_{33}^l \sim k_{33}^t$. The

Table XI. Properties of Cubic Crystals

Crystal	k_{14}	$\varepsilon_{11}^T/\varepsilon_0$	$\varepsilon_{11}^S/\varepsilon_0$	d_{14}, pC/N	e_{14}, C/m²	c_{11} (10^{10} N/m²)	c_{12} (10^{10} N/m²)	c_{44}^E (10^{10} N/m²)	c_{44}^D (10^{10} N/m²)	Density, 10^3 kg/m³
ZnS[43]	0.080	8.37	8.32	3.18	0.15	10.5	6.5	4.61	4.64	4.09
ZnSe[43]	0.026	9.12	9.12	1.10	0.049	8.1	4.9	4.41	4.41	5.26
ZnTe[43]	0.017	10.1	10.1	0.91	0.028	7.1	4.1	3.12	3.12	5.64
CdTe(−196°C)[43]	0.026	9.65	9.65	1.68	0.034	6.2	4.3	1.96	1.96	5.84
GaAs[48,49]	0.065	11.1	11.1	2.60	0.15	11.9	5.38	5.94	5.96	5.31
$Bi_{12}GeO_{20}$[50]	0.266	41	38	31	0.71	12.0	3.9	2.3	2.5	9.20
CuCl[51]	~0.12	~7.5	—	—	—	—	—	—	—	4.14

Table XII. Properties of $LiTaO_3$ and $LiNbO_3$ (3 *m*)

Material	k_{33}^l	k_{31}^l	"k"$_{15}$ ^{a}	k_{33}^t	$\varepsilon_{33}^T/\varepsilon_0$	$\varepsilon_{33}^S/\varepsilon_0$	$\varepsilon_{11}^T/\varepsilon_0$	$\varepsilon_{11}^S/\varepsilon_0$
$LiNbO_3$	0.16	−0.03	0.61	0.16	30	29	84	44
$LiTaO_3$	0.18	−0.05	0.41	0.18	45	43	51	41

Table XII (*continued*)

Material	d_{33}	d_{31}	d_{22}	d_{15}	e_{33}	e_{31}	e_{22}	e_{15}
	pC/N				C/m²			
$LiNbO_3$	6	−1	21	68	1.3	0.2	2.5	3.7
$LiTaO_3$	8	−2	7	26	1.9	0.0	1.6	2.6

Material	s^E_{11}	s^E_{33}	s^E_{12}	s^E_{13}	s^E_{14}	s^E_{44}	s^E_{66}
	pm²/N						
$LiNbO_3$	5.78	5.02	−1.01	−1.47	−1.02	17.0	13.6
$LiTaO_3$	4.87	4.36	−0.58	−1.25	0.64	10.8	10.9

Material	c^E_{11}	c^E_{33}	c^E_{12}	c^E_{13}	c^E_{14}	c^E_{44}	c^E_{66}	c^D_{33}	c^D_{44}	Density, 10^3 kg/m³
	10^{10} N/m²									
$LiNbO_3$	20.3	24.5	5.3	7.5	0.9	6.0	7.5	25.2	9.5	4.7
$LiTaO_3$	23.3	27.5	4.7	8.0	−1.1	9.4	9.3	28.4	11.3	5.3

Note: Curie points—$LiNbO_3 \sim 1200°C$, $LiTaO_3 \sim 660°C$.

[a] Not a pure mode; "k"$_{15} = e_{15}/\sqrt{c^D_{44}\varepsilon^S_{11}}$.

elastic constants and coupling factors change relatively little with temperature below about 10°C, but as is characteristic with ferroelectrics, d_{33} and ε_{33} change drastically. Of all known piezoelectrics, SbSI has by far the strongest response to hydrostatic stress and maintains the highest level of piezoelectric coupling.

With the increasing emphasis on high-frequency transducers interest in piezoelectric crystals with significant conductivity has increased. Examples are gallium arsenide (GaAs), selenium (Se), and tellurium (Te). Se and Te are of the symmetry class of quartz, 32. Coupling of over 0.3 was reported for tellurium,[58] although conductivity was excessive even at 77°K. Data obtained for selenium[59] indicate that $k_{11}^{t} \sim 0.25$, and that conductivity is considerably lower.

7. PARAMETERS OF PIEZOELECTRIC CERAMICS

7.1. General

Piezoelectric ceramics have been optimized for specific applications in large variety by compositional adjustment. By far the most widely used compositions are based on lead titanate–lead zirconate solid solutions.[60] Most of these are close to the phase boundary between a titanium-rich tetragonal phase and a zirconium-rich rhombohedral phase (Fig. 11). Piezoelectric coupling, permittivity, and compliance have maximum values at the phase boundary, but the piezoelectric coupling remains fairly high over a wide range on the zirconium side.[61] All lead titanate zirconate compositions listed in Table XIII, except PZT-2, PZT-7A, and PZT-6B, are just on the tetragonal side of the phase boundary (Fig. 11). PZT-2 and PZT-7A are rhombohedral and PZT-6B is well into the tetragonal field.

Calcium, strontium,[65] or barium[66] may be substituted for a fraction of the lead, and tin may be used for zirconium,[67] resulting in lowered Curie point and increased permittivity (PZT-4). In these cases the atom substituted is isovalant with the replaced atom. When the substituted atom has higher or lower valence more profound changes occur. The electron-donor-doped lead titanate zirconates with, for instance, Nb^{5+} replacing $(Ti, Zr)^{4+}$, or La^{3+} replacing Pb^{2+}, are characterized by enhanced piezoelectric coupling, permittivity, and compliance with increased tan δ_E and tan δ_M, increased dc resistivity, and reduced aging rates[68] (PZT-5 series). The electron-acceptor-doped lead titanate zirconates, for instance with Sc^{3+} or Fe^{3+} replacing $(Ti, Zr)^{4+}$, are characterized by decreased piezoelectric coupling, permittivity, compliance, tan δ_E, and tan δ_M. Aging rates are not significantly affected and dc resistivity decreases slightly (PZT-8). Most effects of donor and acceptor doping are attributed to the types of lattice vacancies that arise. With the former vacancies occur in Pb sites, one vacancy for every two donor atoms substituted. With acceptor doping vacancies occur at the oxygen lattice sites, one vacancy for every two acceptor atoms substituted. With Pb-site vacancies domain-wall motion occurs more readily, and the reversible part of this contributes lossy additional components to the permittivity, compliance, and piezoelectric effect. The ready relaxation of poling strains relieves internal stress and reduces aging effects. With oxygen vacancies domain-wall motion is inhibited, accounting for the opposite effects. Recognition that undoped lead titanate zirconate is a *p*-type conductor accounts for the opposite effects of donor and acceptor doping on volume resistivity.[69,70]

Table XIII lists sets of constants for commonly used piezoelectric ceramics, which are self-consistent insofar as possible. As is stated in the 1961 *IRE Standards on Piezoelectric Crystals*,[3] however, the dielectric, elastic, and piezoelectric constants for a given composition may vary as much as 20, 5, and 10%, respectively, due to fluctuations related to raw material and manufacturing variations.

Table XIV lists typical applications of the piezoelectric ceramics together with the feature which makes each composition particularly suited to its specialized applications. The leading position of the lead titanate zirconates is immediately evident. This is ascribed to their intrinsically strong piezoelectric effects and high Curie points, which allow a wide variation in chemical composition to obtain a large variety of operating characteristics without serious reduction of the piezoelectric effect.

The PZT-6 and PZT-8 compositions have features particularly noteworthy. The latter has remarkably low dielectric and elastic losses at high drive levels. The PZT-6 compositions have temperature and time stability of resonance frequency sufficient for use in voice channel intermediate-frequency filtering in the range from about 200 kHz to a few MHz. Narrower band filters utilizing flexural modes are also used. The bandwidths of ceramic filters are controlled by adjustment of poling level, k_p from 0.20 to 0.44 for PZT-6A and from 0.0 to 0.25 for PZT-6B.

Figures 15 through 18 show changes of parameters with temperature for several lead titanate zirconate compositions. The remarkable differences result from changes in the atom ratio Zr/Ti, isomorphous substitution, and control of donor–acceptor balance. The effects of the transition-metal additives (PZT-6 and PZT-7A) are profound but little understood.

Table XIII.A. Properties of Commonly Used Piezoelectric Ceramics–Low Signal Values at 25°C

Material	k_p	k_{31}^l	k_{33}^l	k_{15}	k_{33}^t	$\varepsilon_{33}^T/\varepsilon_0$	Q_E	$\varepsilon_{33}^S/\varepsilon_0$	$\varepsilon_{11}^T/\varepsilon_0$	$\varepsilon_{11}^S/\varepsilon_0$
PZT-4[a]	−0.58	−0.33	0.70	0.71	0.51	1300	250	635	1475	730
PZT-5A[a]	−0.60	−0.34	0.705	0.685	0.49	1700	50	830	1730	916
PZT-5H[a]	−0.65	−0.39	0.75	0.675	0.505	3400	50	1470	3130	1700
PZT-6A[a]	−0.42	−0.25	0.54	—	0.39	1050	50	730	—	—
PZT-6B[a]	−0.25	−0.145	0.375	0.377	0.30	460	110	386	475	407
PZT-7A[a]	−0.51	−0.30	0.66	0.67	0.50	425	60	235	840	460
PZT-8[a]	−0.51	−0.30	0.64	0.55	0.48	1000	250	580	1290	900
PZT-2[a]	−0.47	−0.28	0.63	0.70	0.51	450	200	260	990	504
$BaTiO_3$	−0.36	−0.21	0.50	0.48	0.38	1700	100	1260	1450	1115
95w% $BaTiO_3$, 5w% $CaTiO_3$	−0.33	−0.19	0.48	0.48	0.38	1200	170	910	1300	1000
NRE-4(62)[b]	−0.31	−0.18	0.46	0.46	0.36	1420	200	1110	—	—
$PbNb_2O_6$[c]	−0.07	−0.045	0.38	—	0.37	225	100	190	—	—
$Pb_{0.6}Ba_{0.4}Nb_2O_6$(63)	−0.38	−0.22	0.55	—	—	1500	100	—	—	—
$Na_{0.5}K_{0.5}NbO_3$(64)	−0.46	−0.27	0.605	0.645	0.46	496	70	306	938	545

Table XIII.A (*continued*)

Material	d_{33}	d_{31}	d_{15}	g_{33}	g_{31}	g_{15}	e_{33}	e_{31}	e_{15}
		pC/N			10^{-3} Vm/N			C/m^2	
PZT-4[a]	289	−123	496	25.1	−10.7	38.0	15.1	−5.2	12.7
PZT-5A[a]	374	−171	584	24.9	−11.4	38.0	15.8	−5.4	12.3
PZT-5H[a]	593	−274	741	19.7	−9.1	26.8	23.3	−6.5	17.0
PZT-6A[a]	189	−80	—	20.4	−8.6	—	12.5	—	—
PZT-6B[a]	71	−27	130	17.4	−6.6	31.0	7.1	−0.9	4.6
PZT-7A[2]	150	−60	362	39.8	−15.9	48.8	9.5	−2.1	9.2
PZT-8[a]	225	−97	330	25.4	−10.9	29.0	13.2	−4.0	10.4
PZT-2[a]	152	−60	440	38.2	−15.1	50.1	9.0	−1.9	9.8
$BaTiO_3$	190	−78	260	12.6	−5.2	20.2	17.5	−4.3	11.4
95w% $BaTiO_3$, 5w% $CaTiO_3$	149	−58	242	14.0	−5.45	21.0	13.5	−3.1	10.9
NRE-4[(62)b]	150	−59	—	11.9	−4.7	—	—	—	—
$PbNb_2O_6$[c]	85	∼−9	—	42.5	∼−4.5	—	—	—	—
$Pb_{0.6}Ba_{0.4}Nb_2O_6$[(63)]	220	−90	—	16.6	−6.8	—	—	—	—
$Na_{0.5}K_{0.5}NbO_3$[(64)]	127	−51	306	29.0	−11.6	36.9	9.8	—	11.3

[a] Trademark, Vernitron Piezoelectric Division.
[b] 95 w% $BaTiO_3$, 5 w% $CaTiO_3$, plus 0.75 w% $CoCO_3$.
[c] General Electric Company.

Table XIII.B. Properties of Commonly Used Piezoelectric Ceramics–Low Signal Values at 25°C

Material	s_{33}^E	s_{11}^E	Q_M	s_{44}^E	s_{66}	s_{33}^D	s_{11}^D	s_{44}^D	c_{33}^E	c_{11}^E	c_{33}^D	c_{11}^D	N_1	N_{3t}	v_3^D	v_4^D	Density, 10^3 kg/m^3
	pm^2/N								10^{10} N/m^2				Hz · m		m/sec		
PZT-4	15.5	12.3	500	39.0	32.7	7.90	10.9	19.3	11.5	13.9	15.9	14.5	1650	2000	4600	2630	7.5
PZT-5A	18.8	16.4	75	47.5	44.3	9.46	14.4	25.2	11.1	12.1	14.7	12.6	1400	1890	4350	2260	7.75
PZT-5H	20.7	16.5	65	43.5	42.6	8.99	14.1	23.7	11.7	12.6	15.7	13.0	1420	2000	4560	2375	7.5
PZT-6A	13.0	10.7	450	—	27.8	9.2	10.1	—	13.1	—	15.5	—	1770	2140	4570	—	7.45
PZT-6B	9.35	9.0	1300	28.2	24.0	8.05	8.8	24.2	16.3	16.8	17.7	16.9	1920	2225	4820	2340	7.55
PZT-7A	13.9	10.7	600	39.5	27.8	7.85	9.7	21.8	13.1	14.8	17.5	15.7	1750	2100	4800	2490	7.6
PZT-8	13.5	11.5	1000	31.9	29.8	8.0	10.4	22.6	12.3	13.7	16.1	14.0	1700	2070	4580	2420	7.6
PZT-2	14.8	11.6	680	45.0	29.9	9.0	10.7	22.9	11.3	13.5	14.8	13.6	1680	2090	4410	2400	7.6
$BaTiO_3$	9.5	9.1	300	22.8	23.6	7.1	8.7	17.5	14.6	15.0	17.1	15.0	2200	2520	5470	3160	5.7
95w% $BaTiO_3$, 5w% $CaTiO_3$	9.1	8.6	400	22.2	22.4	7.0	8.3	17.1	15.0	15.8	17.7	15.9	2290	2740	5640	3240	5.55
NRE-4	—	8.1	—	—	—	—	—	—	—	—	—	—	2310	2760	—	—	5.7
$PbNb_2O_6$	25.4	—	11	—	—	21.8	—	—	—	—	—	—	—	—	—	—	6.0
$Pb_{0.6}Ba_{0.4}Nb_2O_6$	—	11.5	250	—	—	—	10.9	—	—	—	—	—	1915	—	—	—	5.9
$Na_{0.5}K_{0.5}NbO_3$	10.1	8.2	240	27.0	—	6.4	7.6	15.8	16.8	—	21.4	—	2570	—	6940	3760	4.46

Table XIII.C. Properties of Commonly used Piezoelectric Ceramics

Material	Curie point, °C	Heat capacity, J/kg °C	Thermal cond., W/m °C	Static tensile strength, psi	Rated dynamic tensile strength, psi	Change in N_1 per time decade, %	Change in k_p per time decade, %	Change in ε_{33}^T per time decade, %	Change in N_1 −60 to +85°C, %
PZT-4	328	420	2.1	11000	6000	+1.5	−2.3	−5.8	4.8
PZT-5A	365	420	1.5	11000	4000	+0.2	−0.2	−1.0	2.6
PZT-5H	193	420[a]	1.5[a]	11000[a]	4000[a]	+0.25	−0.35	−1.5	9.0
PZT-6A	335	420[a]	2.1[a]	11000[a]	4000[a]	<0.1	−0.2	−0.6	<0.2
PZT-6B	∼350	420[a]	2.1[a]	11000[a]	3500[a]	<0.1	−0.2	−0.6	<0.2
PZT-7A	∼350	420[a]	2.1[a]	11000[a]	3500[a]	−0.08	0.0	+2.0	2.9
PZT-8	300	420[a]	2.1[a]	12000	7000	+1.0	−2.0	−5.0	∼2.0
PZT-2	370	420[a]	2.1[a]	11000[a]	3500[a]	+0.6	−1.8	−2.8	1.5
$BaTiO_3$	115	500	3.5	7500	3000	+1.1	−2.5	−4.1	19.0
95w% $BaTiO_3$, 5w% $CaTiO_3$	115	500[a]	3.5[a]	7500[a]	3000[a]	+0.5	−1.8	−0.8	∼18
NRE-4	105	500[a]	3.5[a]	7500[a]	3000[a]	+0.4	−1.9	−1.3	∼15
$PbNb_2O_6$	570	—	—	2500[a]	—	—	—	—	—
$Pb_{0.6}Ba_{0.4}Nb_2O_6$	260	—	—	—	—	—	—	—	—
$Na_{0.5}K_{0.5}NbO_3$	420	—	—	—	—	—	—	—	—

Table XIII.C (*continued*)

Material	Dielectric time constants, s: 25°C	Dielectric time constants, s: 100°C	Dielectric time constants, s: 200°C	ac depoling field 25°C, kV/cm rms	ac field in kV/cm rms for tan $\delta_E = 0.04$: 25°C	ac field in kV/cm rms for tan $\delta_E = 0.04$: 100°C	Polarization, 10^{-2} C/m^2
PZT-4	> 100	~5	~0.07	>10	3.9 (17%)[b]	3.3	30
PZT-5A	>2000	~1800	~250	~7	0.45 (11%)[b]	0.45	38
PZT-5H	>2000	>2000	~1000	~4	0.3 (5%)[b]	0.2	33
PZT-6A	>10	~0.15	~0.002	~8	3.2	2.8	30
PZT-6B	>100	~1.5	~0.03	>10	11.0	5.0	15
PZT-7A	>10	~0.5	~0.03	>10	2.6	1.3	42
PZT-8	>100	~2	~0.01	>15	tan $\delta_E = 0.015$ @ 6 kV/cm(8%)	tan $\delta_E = 0.03$ @ 6 kV/cm	25
PZT-2	>100	~2	~0.03	>10	1.8	1.6	40
$BaTiO_3$	>150	~0.5	~0.002	~4	1.0	0.8 (75°C)	8
95w% $BaTiO_3$, 5w% $CaTiO_3$	>100	~0.3	~0.002	~4	1.7	1.0 (75°C)	8
NRE-4	—	—	—	~9	~4.0	—	8
$PbNb_2O_6$	>1000	~500	~10	>10	>6	—	—
$Pb_{0.6}Ba_{0.4}Nb_2O_6$	>1000	~50	—	—	1.4 (17%)[b]	—	—
$Na_{0.5}K_{0.5}NbO_3$	~50	—	—	—	—	—	~30

[a] Estimated.
[b] The number in parenthesis gives increase in ε_{33}^T at the indicated electric field.

Table XIV. Applications of Piezoelectric Ceramic Compositions

Material	Application	Outstanding features
PZT-4	Sonar radiators, ultrasonic transducers, high-voltage generators	Good high-drive characteristics, high coupling
PZT-5A, PZT-5H	Hydrophones, instrument transducers, phonograph pickups, microphone and loudspeaker elements	High permittivity and coupling, low aging, high time constant
PZT-6A, PZT-6B	Electric wave filters	Outstanding temperature stability, low aging
PZT-7A	Delay line transducers	Low permittivity, low aging
PZT-8	Sonar radiators, ultrasonic transducers	Outstanding high drive characteristics
$BaTiO_3$	Sonar radiators, ultrasonic transducers	Excessive temperature dependence
95 w% $BaTiO_3$, 5 w% $CaTiO_3$	Sonar radiators, ultrasonic transducers	Low cost, reasonable stability at low power levels
95 w% $BaTiO_3$, 5 w% $CaTiO_3$, 0.75 w% $CoCO_3$-NRE-4	Sonar radiators	Good high-drive characteristics but weak piezoelectric effect
$PbNb_2O_6$	Ultrasonic flaw detection transducer	Low mechanical Q
$Pb_{0.6}Ba_{0.4}Nb_2O_6$		No outstanding features
$Na_{0.5}K_{0.5}NbO_3$ (hot-pressed)	Delay line transducers	Low permittivity, high wave velocity

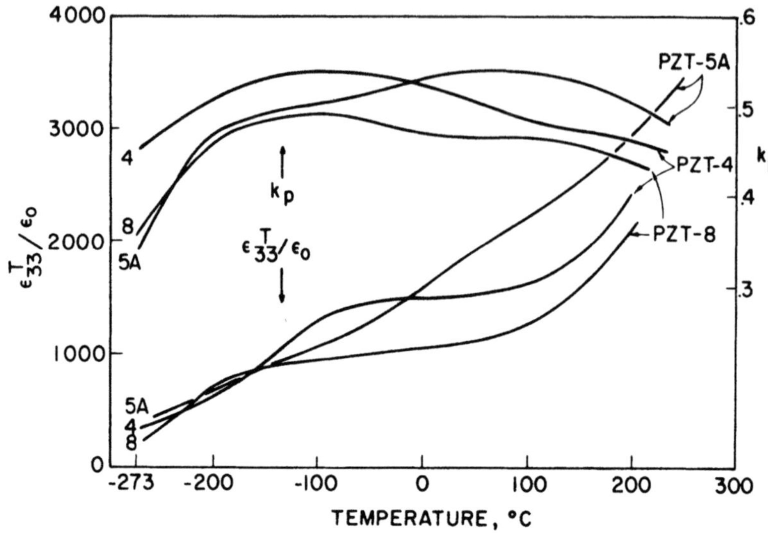

Fig. 15. Permittivity (ε^T_{33}) and k_p *vs* temperature for temperature-stabilized PZT-4, PZT-8, and PZT-5A.

Unique compositions not in the lead titanate zirconate family are $PbNb_2O_6$ and $Na_{.5}K_{.5}NbO_3$. The former has a remarkedly high mechanical damping and a strong response to hydrostatic pressure. The mechanical damping, which suppresses ringing, is particularly useful in ultrasonic flaw detection. Modified lead metaniobate compositions have not found

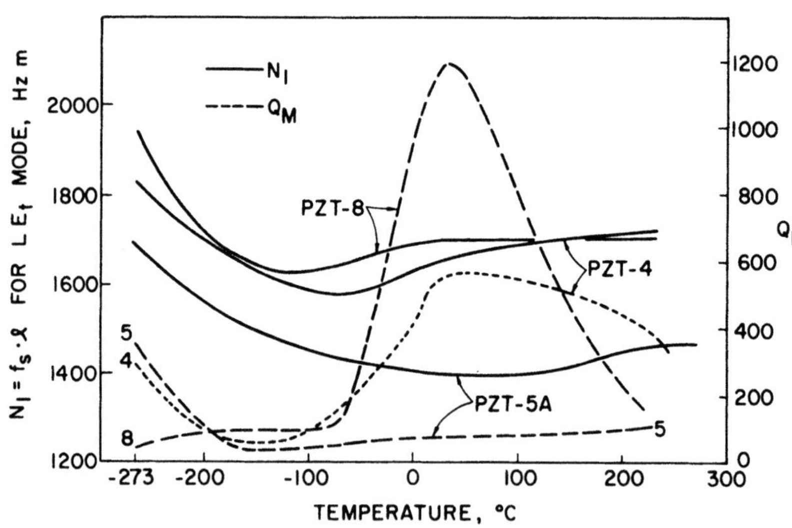

Fig. 16. Q_M and N_1 *vs* temperature for temperature-stabilized PZT-4, PZT-8, and PZT-5A.

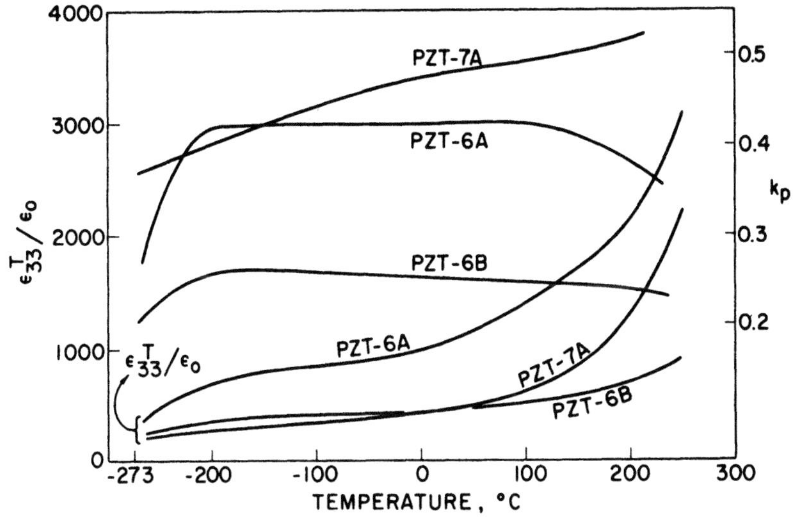

Fig. 17. Permittivity (ε_{33}^T) and k_p vs temperature for temperature-stabilized PZT-6A, PZT-6B, and PZT-7A.

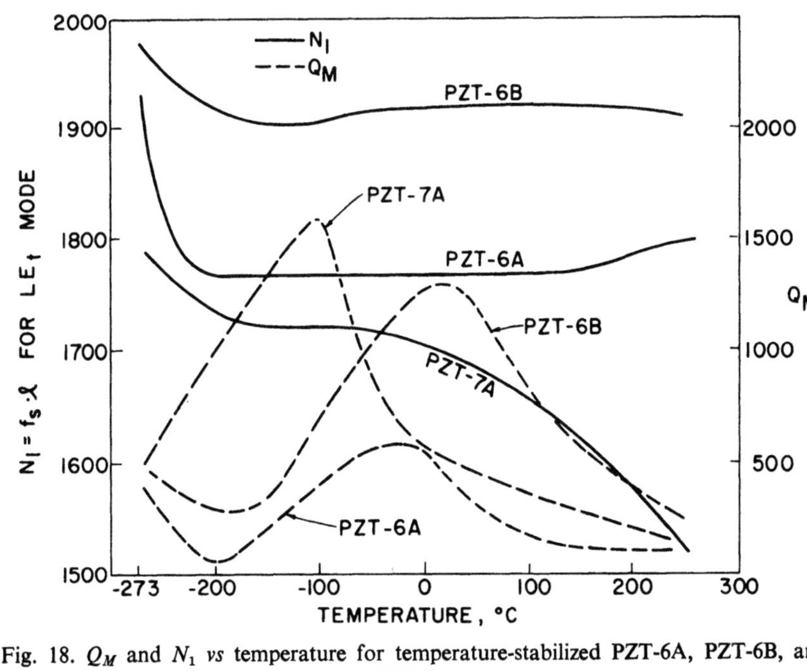

Fig. 18. Q_M and N_1 vs temperature for temperature-stabilized PZT-6A, PZT-6B, and PZT-7A.

significant application. The major application of $Na_{.5}K_{.5}NbO_3$ in high-frequency transducers (10–200 MHz) is ascribed to its low permittivity and high elastic wave velocity.

7.2. Aging in Piezoelectric Ceramics and Effects of High Static Stress

The dielectric and elastic losses ascribed respectively to high-amplitude electric field and dynamic stress were discussed in Section 5.1 and nonlinearities were discussed in general terms in Section 4.3. In this section additional effects which also may be generally ascribed to the ferroelectric nature of these materials, i.e., domain-wall motion, are summarized.

So long as the electric field and mechanical stress are of very low amplitude the piezoelectric ceramics may be considered linear. Thresholds for deviation from linearity vary widely for the various piezoelectric ceramics, as discussed previously. If a static stress or electric field is maintained and periodic driving stress or electric field is superposed, additional domain-wall motion may be encouraged or constrained. The wide variety of possible operating conditions and domain-wall effects make comprehensive characterization of the piezoelectric ceramics difficult.

General principles and normal operating conditions are therefore emphasized in this section. Losses were covered generally in Section 5. Another important factor in use of piezoelectric ceramics is the stability of the parameters under widely varying operating characteristics. Sonar transducers are in many cases exposed to high mechanical stress during deep submergence. Radiating transducers are also exposed to high-amplitude dynamic stress, and in low Q transducers electric fields are high. There are, in addition, many cases with fairly severe thermal cycles.

7.2.1. General Discussion of Aging and Stability

Aging of ferroelectric materials is roughly logarithmic in nature, with properties varying with time after poling as sketched in Fig. 19. The rates shown are typical for ceramics used in radiating transducers. Aging rates are only one-fifth to one-tenth as great with the donor-doped compositions whose poor high-drive characteristics exclude them, however, from use in radiating transducers. There are obvious limitations to the logarithmic aging characterization, but it usually holds fairly well in the range one to 50 weeks after poling. After about a year there is typically a slight decrease in rate of change per decade of time.

Exposure to a high mechanical stress, a high electric field, or an elevated temperature at some point along the aging curve creates in general the

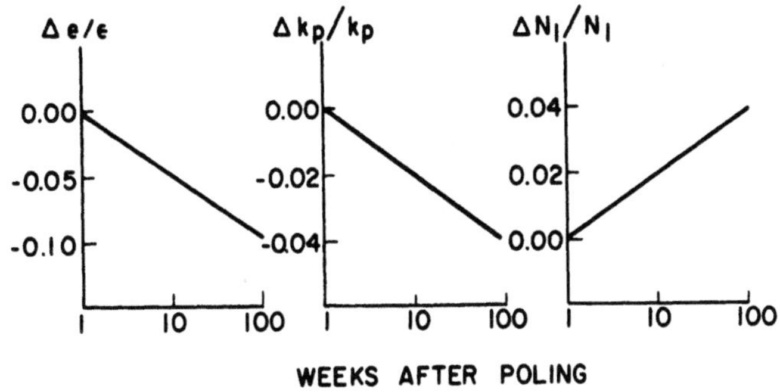

Fig. 19. Aging curves (ε_{33}^T, k_p, and N_1) for PZT-8.

effects sketched in Fig. 20. One may interpret the change in each parameter as consisting of "permanent" (Δp) and "temporary" (Δt) components, as shown in Fig. 20. The permanent and temporary changes have the same sign (positive) for permittivity. With piezoelectric coupling the permanent change is usually negative, the temporary change always positive. The permanent and temporary changes of frequency constant are both negative. The amount of temporary change occurring at any disturbance increases with the time elapsed between poling or the last major disturbance and the new disturbance. If one takes the major disturbance as the initiation of a new aging cycle and replots on this basis, with a logarithmic time scale, a linear plot is again obtained. One can, therefore, consider exposure to high temperature, pressure, or electric field or any combination thereof as the initiation of a new aging cycle.

The permanent set may for some materials and certain disturbances take place only once, and in such case the disturbance acts to "stabilize."

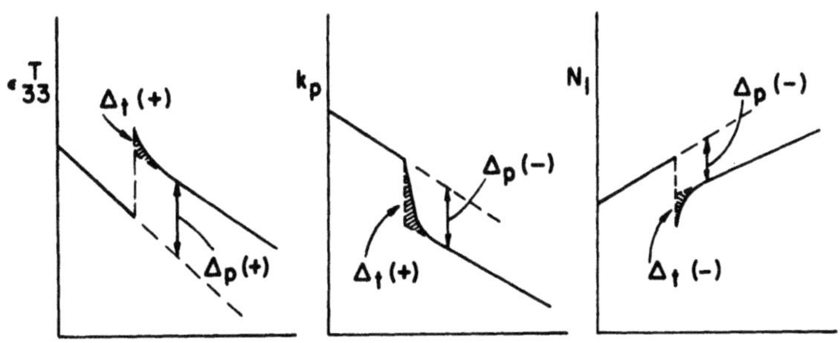

Fig. 20. Aging curves showing effects of a major disturbance (exposure to high electric field, stress, or temperature). The signs of the "permanent" (Δp) and "temporary" (Δt) changes are shown.

Heat soak has been used to eliminate for future heat cycles the permanent change with PZT-5A. It has been demonstrated that a heat soak not only can eliminate the permanent change with PZT-4 for future heat cycles, but can reduce drastically the permanent and temporary changes with pressure.[71,72] Ideally these stabilizing treatments should be provided shortly after poling in order that the newly initiated aging cycle date back to the earliest time.

The temporary changes are in the nature of "deaging," since they are positive for piezoelectric coupling and permittivity, while negative for frequency constant. The temporary changes are less on freshly poled specimens and greater on well-aged specimens. The magnitudes of the permanent changes depend little upon how well-aged the specimens are.

The total changes ($\Delta p + \Delta t$) connected with a disturbance are roughly proportional to the magnitude of the disturbances, but markedly different for different compositions. Furthermore, there is a fundamental difference between donor-doped on the one hand and neutral or acceptor-doped lead titanate zirconate ceramics on the other hand with respect to the cumulative permanent effect of repeated disturbances. An outstanding example of cumulative loss results from exposure of donor-substituted lead titanate zirconate to repeated parallel stress cycles in the 10 kpsi range.[71,72] A single cycle results in loss of only a few percent of the remanent polarization, but additional losses during subsequent cycles lead to severe reduction in the piezoelectric effect (note Fig. 24). The temporary component of the disturbance with these materials is relatively small. Since this is actually a deaging effect, the subsequent aging is essentially similar to that immediately after poling.

With the neutral or acceptor-substituted lead titanate zirconates repeated disturbances fortunately have relatively little progressive effect (see Fig. 23). If a considerable time has elapsed since the last major disturbance, however, a repetition of the identical disturbance again creates a temporary but not a permanent change. The sketches in Fig. 21 illustrate these opposing features of the neutral or acceptor-modified and the donor-modified lead titanate zirconates for a repeated major stress disturbance. It is readily seen that the latter materials cannot be applied where major stress disturbances occur repeatedly. The general features of this discussion are summarized below:

1. Total changes upon return to ambient conditions immediately after a given disturbance are usually least when the changes upon going from ambient to the disturbed condition are small.

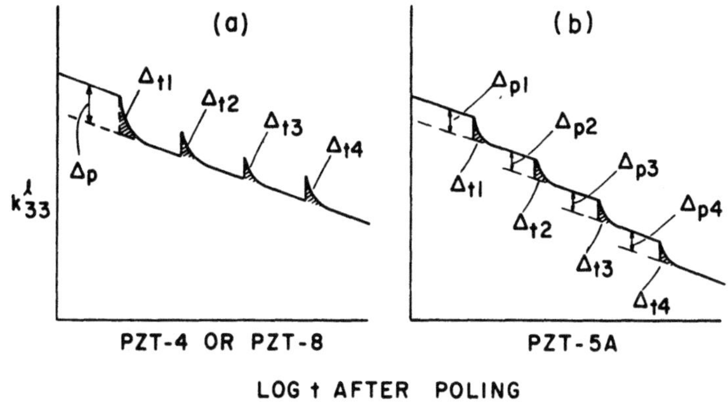

Fig. 21. Aging curves for k_{33}^l showing effects of major stress disturbances on a) the neutral or acceptor-doped lead titanate zirconates and b) the donor-doped lead titanate zirconates.

2. Temporary changes are small when ambient aging rates are low.
3. Permanent changes are small when the disturbance does not have a major effect on domain orientation. Materials with low permanent changes are therefore those which exhibit good high-drive characteristics.
4. All presently available piezoelectric transducer materials which have small temporary changes (low aging materials) have poor high-drive characteristics.

7.2.2. Effects of High Static Stress

The effects of high static stress on operating characteristics of lead titanate zirconate ceramics have been thoroughly studied.[71–80] The overall picture is extremely complicated, however, and this discussion will, therefore, be fairly general. Static stress configurations which may be encountered are listed below:

1. One-dimensional compression parallel to the polar axis.
2. One-dimensional compression perpendicular to the polar axis.
3. Two-dimensional compression perpendicular to the polar axis.
4. Hydrostatic compression.

Stress cannot cause 180° domain reorientation, and therefore all effects of static stress are due to domain reorientation other than 180° (90° in tetragonal and 71°/109° in rhombohedral perovskites).

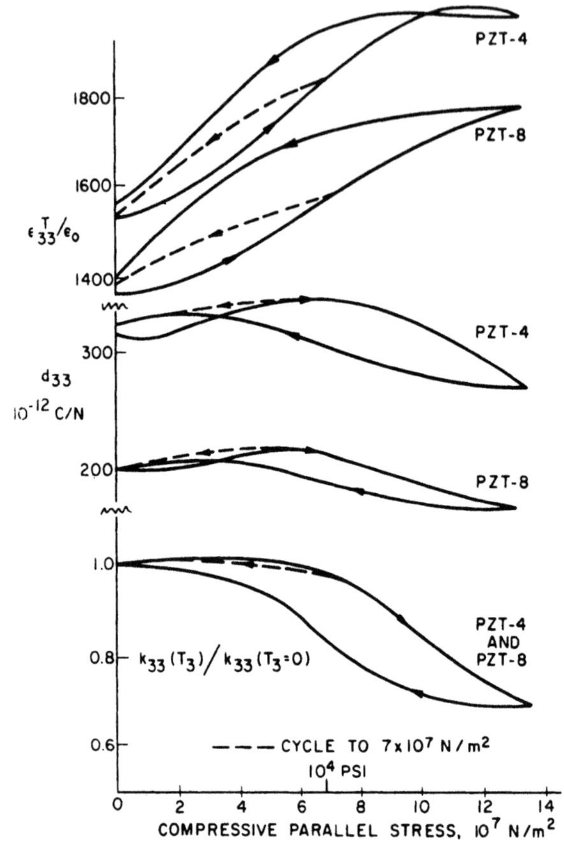

Fig. 22. Permittivity (ε_{33}^{T}), d_{33}, and k_{33}^{l} *vs* compressive parallel stress for temperature-stabilized PZT-4 and PZT-8.

Figure 22 shows the changes of the free permittivity, d_{33} and k_{33}^{l} with stress parallel to the polar axis for temperature-stabilized PZT-4 and PZT-8.* It is noteworthy that k_{33}^{l} is virtually unchanged in the range to 7×10^7 N/m^2 ($\sim 10^4$ psi) for both materials and that even though k_{33}^{l} decreases by about one-third at 14×10^7 N/m^2 ($\sim 2 \times 10^4$ psi), there is virtually no permanent change.

Figure 23 shows permittivity and d_{33} as functions of the number of parallel stress cycles to 7×10^7 N/m^2 ($\sim 10^4$ psi) for PZT-4 and PZT-8.

* The temperature stabilization consists of a one-hour soak at 200°C. This causes relatively little degradation of the piezoelectric characteristics, but significantly reduces changes of parameters with pressure and subsequent temperature cycling.

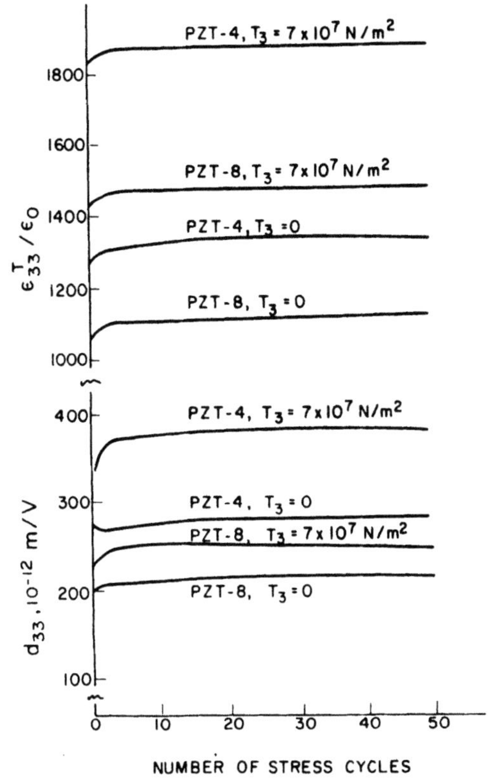

Fig. 23. Permittivity (ε_{33}^T) and d_{33} *vs* number of parallel compressive stress cycles for unstabilized PZT-4 and PZT-8.

Values measured with stress maintained ($T_3 = 7 \times 10^7$ N/m^2) and immediately after stress release ($T_3 = 0$) are shown. Note that stress stabilization occurs and is virtually complete after about five repetitive cycles. Temperature stabilization to decrease effects of static mechanical stress is permanent; stabilization introduced by stress cycling, however, is lost in subsequent aging.

Figure 24 demonstrates that the characteristics for PZT-5H are drastically changed by exposure to 7×10^7 N/m^2 ($\sim 10^4$ psi). Similar studies indicate that stress cycling should be limited for PZT-5H to about 2×10^7 N/m^2 ($\sim$2500 psi) parallel stress.

Figure 25 shows normalized ε_{33}^T and tan δ_E *vs* rms electric field for PZT-4 and PZT-8 with and without parallel bias stress. The deaging effect of the static stress clearly acts to increase domain mobility and drastically in-

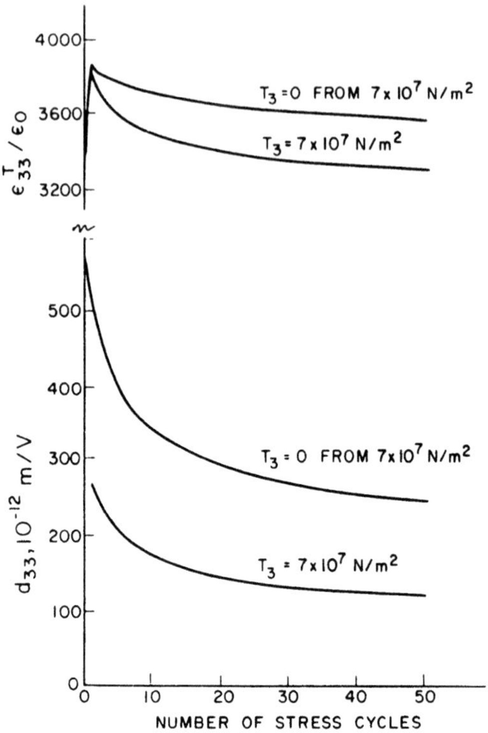

Fig. 24. Permittivity (ε_{33}^T) and d_{33} *vs* number of parallel compressive stress cycles for unstabilized PZT-5H.

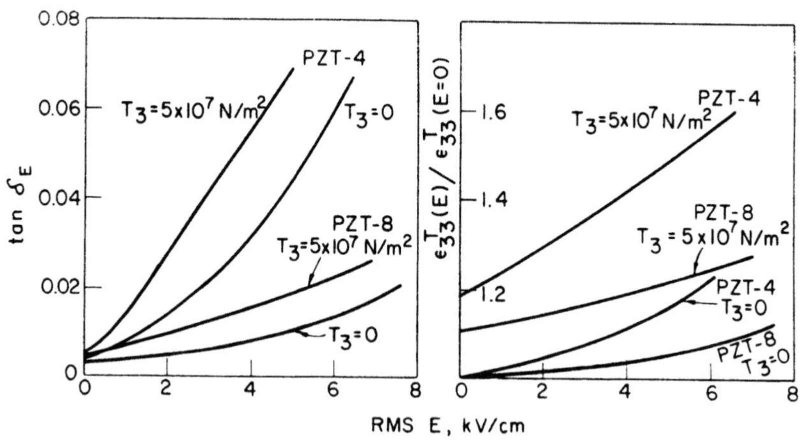

Fig. 25. Permittivity (ε_{33}^T) and tan δ_E *vs* electric field with and without parallel compressive stress of 5×10^7 N/m² for unstabilized PZT-4 and PZT-8.

crease losses. The effects of the acceptor-doping in PZT-8 are particularly helpful here.

Figure 26 shows effects of one-dimensional compressive stress perpendicular to the polar axis for PZT-4 and PZT-8. Changes of ε_{33}^T are smaller than with parallel stress. Stress, however, serves to destroy the planar symmetry of the poled ceramic elements. The effect, unfortunately, is to decrease the piezoelectric constant (d_{31}) that produces strain in the direction of the applied stress, i.e., the loaded direction. The piezoelectric constant (d_{32}) which produces strain perpendicular to the static stress and the polar axis is increased; $(d_{31} + d_{32})/2$ with stress maintained is about equal to the initial value of d_{31}. There is very little permanent reduction of symmetry when the applied stress is removed.

Figure 27 summarizes effects of hydrostatic stress, such as free-flooded sonar transducers experience, to well beyond ocean pressure. This is by far the most favorable stress configuration for high stability. Nevertheless,

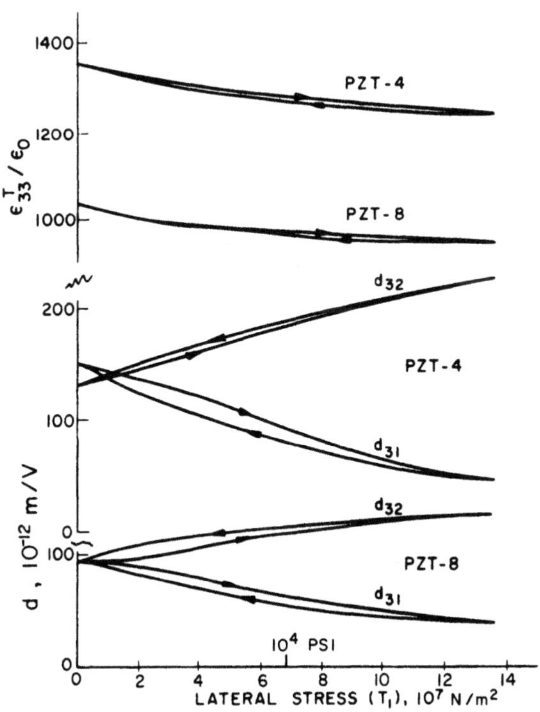

Fig. 26. Permittivity (ε_{33}^T) and lateral piezoelectric constants *vs* lateral compressive stress for unstabilized PZT-4 and PZT-8.

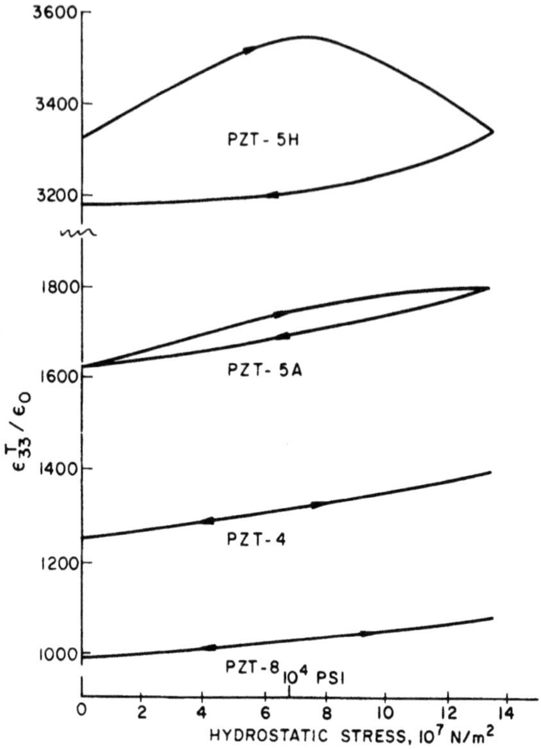

Fig. 27. Permittivity (ε_{33}^T) *vs* hydrostatic stress for several piezoelectric ceramics (unstabilized).

severe hysteresis exists with the donor-doped compositions. With PZT-5A, however, hysteresis is negligible below 7×10^7 N/m^2. Changes in piezoelectric constants are not shown in Fig. 27; these are very small for the neutral and acceptor-doped compositions and also for PZT-5A with pressure below 7×10^7 N/m^2.

Figure 28 shows effects of static planar stress for PZT-4. This is the stress configuration experienced by a spherical shell, with $T_p \approx pD/4t$, where D is the diameter, t the wall thickness, and p the hydrostatic pressure. With stress maintained this configuration produces, by far, the most deleterious effects, but there is no tendency for progressive depolarization. Note that the effect of stress cycles to 40 kpsi is to *increase* d_{31} by 17% and ε_{33}^T about 7%. This is a deaging effect.

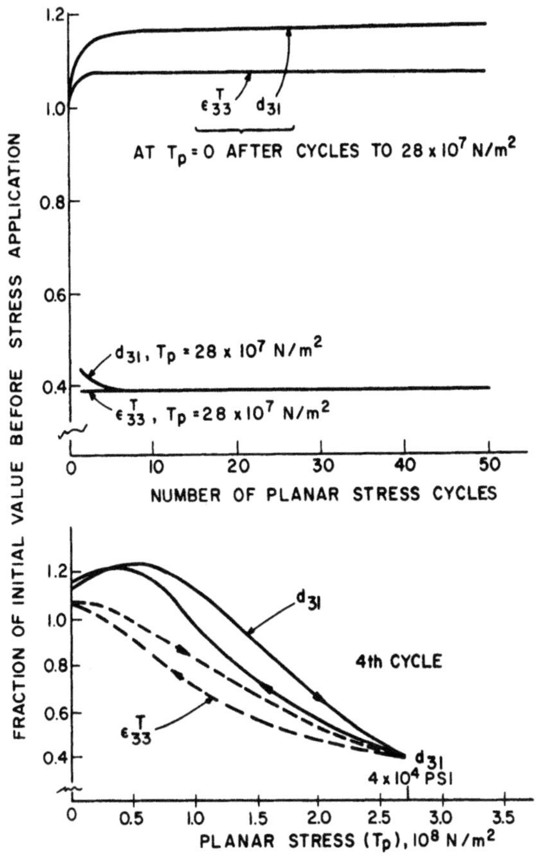

Fig. 28. Effects of planar stress ($T_p = T_1 = T_2$) on d_{31} and ε_{33}^T for unstabilized PZT-4.

NOTATION

E_m	electric field (V/m)
D_m	electric displacement (C/m^2)
P	polarization (C/m^2)
T	stress (N/m^2)
	($T_{ii} = T_i$, $T_{12} = T_6$, $T_{13} = T_5$, $T_{23} = T_4$)
S	strain (m/m)
	($S_{ii} = S_i$, $S_{12} = S_6/2$, $S_{13} = S_5/2$, $S_{23} = S_4/2$)
p	pressure (N/m^2)
V	voltage (V)
I	current (A)
F	force (N)
U	displacement (m)
P	power per unit volume of ceramic (W/m^3)

W	work (J)
e_{mi}	piezoelectric constant (C/m^2)
d_{mi}	piezoelectric constant (m/V) or (C/N)
g_{mi}	piezoelectric constant (Vm/N)
h_{mi}	piezoelectric constant (V/m)
c_{ij}	elastic stiffness constant (N/m^2)
s_{ij}	elastic compliance constant (m^2/N)
ε_{mk}	dielectric permittivity (C/Vm)
β_{mk}	dielectric impermeability (Vm/C)
k	piezoelectric coupling factor (dimensionless)
$\sigma_p = -s_{12}/s_{11}$	planar Poissons' ratio (dimensionless)
$\sigma_m = -s_{13}/\sqrt{s_{11}s_{33}}$	meridianal Poissons' ratio (dimensionless)
l	length of a bar or rod (m)
w	width of a bar (m)
t	thickness of a plate, bar or ring (m)
h	axial length of a ring (m)
a	radius of sphere or ring (m)
A	area (m^2)
t	time (sec)
ϱ	density (kg/m^3)
f_s	series resonance frequency (Hz)
f_p	parallel resonance frequency (Hz)
f_r	resonance frequency (Hz); lower zero reactance frequency
f_a	antiresonance frequency (Hz); upper zero reactance frequency
f_m	frequency of minimum impedance (Hz)
f_n	frequency of maximum impedance (Hz)
v	wave velocity (m/sec)
Z_0	mechanical impedance (kg/sec)
N	transducer ratio (N/V)
$\tan \delta_E$	dielectric loss factor (dimensionless)
$\tan \delta_M$	mechanical loss factor (dimensionless)
$\tan \delta_{ME}$	piezoelectric loss factor (dimensionless)
$Q_M = 1/\tan \delta_M$	mechanical quality factor (dimensionless)
$Q_E = 1/\tan \delta_E$	electrical quality factor (dimensionless)
Q_L	loaded Q of a transducer (dimensionless)
N_1	$f_s \cdot l$ for LE$_t$ mode of a bar
N_{3t}	$f_s \cdot t$ for TE mode of a plate
k_{mat}	piezoelectric coupling factor for uniform electric field and mechanical stress
k_{eff}	piezoelectric coupling factor for the dynamic case with nonuniform electric field and/or mechanical stress
k_{33}^l	k_{mat} for LE$_p$ mode
k_{31}^l	k_{mat} for LE$_t$ mode
k_p	k_{mat} for PE$_t$ mode
k_{33}^w	k_{mat} for WE$_p$ mode
k_{31}^w	k_{mat} for WE$_t$ mode
k_{36}	k_{mat} for FS$_t$ mode
k_{33}^t	k_{mat} for TE mode

k_{15}	k_{mat} for TS mode
LE_p	longitudinal expander, parallel electric field
LE_t	longitudinal expander, transverse electric field
PE_t	planar expander, transverse electric field
WE_p	width expander, parallel electric field
WE_t	width expander, transverse electric field
FS_t	face shear, transverse electric field
TE	thickness expander, parallel electric field
TS	thickness shear, parallel electric field

REFERENCES

1. "IRE Standards on Piezoelectric Crystals, 1949," *Proc. IRE* **37**, 1378–1395 (1949).
2. "IRE Standards on Piezoelectric Crystals: Determination of the Elastic, Piezoelectric, and Dielectric Constants—The Electromechanical Coupling Factor, 1958," *Proc. IRE* **46**, 764–778 (1958).
3. "IRE Standards on Piezoelectric Crystals: Measurements of Piezoelectric Ceramics, 1961," *Proc. IRE* **49**, 1161–1169 (1961).
4. W. G. Cady, *Piezoelectricity*, Vols. 1 and 2, Dover, New York, 1964 (Revision of McGraw-Hill, New York, 1946).
5. W. P. Mason, *Piezoelectric Crystals and Their Application to Ultrasonics*, Van Nostrand, Princeton, N. J. (1950).
6. W. P. Mason, *Physical Acoustics and the Properties of Solids*, Van Nostrand, Princeton, N. J. (1958).
7. D. Berlincourt, D. R. Curran, and H. Jaffe, "Piezoelectric and Piezomagnetic Materials and their Function in Transducers," in *Physical Acoustics* (W. P. Mason, ed.) Vol. 1A, pp. 169–270, Academic Press, New York (1964).
8. Landolt–Börnstein, "Numerical Data and Functional Relationships in Science and Technology," (K. H. Hellwege, ed.) Group III, Vol. I, (R. Bechmann and R. F. S. Hearmon), Springer, Berlin (1966). See also Vol. II, Group III (1969).
9. H. Jaffe and D. Berlincourt, "Piezoelectric transducer materials," *Proc. IEEE* **53**, 1372–1386 (1965).
10. "Piezoelectric technology data for designers," distributed by Piezoelectric Div., Gould, Inc., Bedford, Ohio (1965).
11. "Piezoelectric ceramics," distributed by Electronic Components and Materials Div., N. V. Philips, Eindhoven, Netherlands (1968).
12. J. F. Nye, *Physical Properties of Crystals*, The Clarendon Press, Oxford, England (1957).
13. D. Berlincourt, "Piezoelectric and ferroelectric energy conversion," *IEEE Trans. on Sonics and Ultrasonics SU-15*, 87–97 (1968).
14. R. Bechmann, "Contour modes of square plates excited piezoelectrically and determination of electric and piezoelectric coefficients," *Proc. Phys. Soc. (London)* **64B**, 323–337 (1951).
15. R. Holland, "Contour extensional resonant properties of rectangular piezoelectric plates," *IEEE Trans. on Sonics and Ultrasonics SU-15*, 97–105 (1968).
16. R. Holland and E. P. Eer Nisse, *Design of Resonant Piezoelectric Devices*, M.I.T. Press, Cambridge, Mass. (1969).

17. W. P. Mason, *Electromechanical Transducers and Wave Filters*, Van Nostrand, New York (1948).
18. W. Shockley, D. R. Curran, and D. J. Koneval, "Trapped-energy modes in quartz filter crystals," *J. Acoust. Soc. Am.* **41**, 981–993 (1967).
19. W. S. Mortley, "Frequency-modulated quartz oscillators for broadcasting equipment," *Proc. IEE* **104B**, 239–249 (1956).
20. R. A. Sykes and W. D. Beaver, "High frequency monolithic crystal filters with possible application to single frequency and single side band use," *Proc. 20th Annual Symposium on Frequency Control*, Atlantic City, N. J. (1966), pp. 288–293.
21. M. Onoe and H. Jumonji, "Analysis of piezoelectric resonators vibrating in trapped-energy modes," *Electronics and Comm. Eng. (Japan)* **48**, 84–93 (1965).
22. H. F. Tiersten, "Thickness vibrations of piezoelectric plates," *J. Acoust. Soc. Am.* **35**, 53–58 (1963).
23. M. Onoe, H. F. Tiersten, and A. H. Meitzler, "Shift in the location of resonant frequencies caused by large electromechanical coupling in thickness mode resonators," *J. Acoust. Soc. Am.* **35**, 36–42 (1963).
24. D. Berlincourt and H. Jaffe, "Elastic and piezoelectric coefficients of single crystal barium titanate," *Phys. Rev.* **111**, 143–148 (1958).
25. D. Berlincourt, H. H. A. Krueger, and B. Jaffe, "Stability of phases in modified lead zirconate with variation in pressure, electric field, temperature, and composition," *J. Phys. Chem. Solids* **25**, 659–674 (1964).
26. D. Berlincourt, "Transducers using forced transitions between ferroelectric and antiferroelectric states," *IEEE Trans. on Sonics and Ultrasonics* **SU-13**, 116–125 (1966).
27. G. E. Martin, *U.S. Navy J. of Underwater Acoustics* **15**, 329–332 (1965).
28. G. E. Martin, "Vibrations of coaxially segmented longitudinally polarized ferroelectric tubes," *J. Acoust. Soc. Am.* **36**, 1496–1506 (1964).
29. R. Holland, "Representation of dielectric, elastic, and piezoelectric losses by complex coefficients," *IEEE Trans. on Sonics and Ultrasonics* **SU-14**, 18–20 (1967).
30. W. P. Mason and R. A. Sykes, "Low-frequency quartz-crystal cuts having low temperature coefficients," *Proc. IRE* **32**, 208–215 (1944).
31. R. C. Miller, "Optical second harmonic generation in piezoelectric crystals," *Appl. Phys. Letters* **5**, 17–19 (1964).
32. K. Nassau, H. J. Levinstein, and G. M. Loiacono, "Ferroelectric lithium niobate. 1. Growth, domain structure, dislocations, and etching," *J. Phys. Chem. Solids* **27**, 983–986 (1966).
33. H. J. Levinstein, A. A. Ballman, and C. D. Capio, "Domain structure and Curie temperature of single crystal lithium tantalate," *J. Appl. Phys.* **37**, 4585–4586 (1966).
34. J. E. Guesic, H. J. Levinstein, J. J. Rubin, S. Singh, and L. G. Van Uitert, "The nonlinear optical properties of $Ba_2NaNb_5O_{15}$," *Appl. Phys. Letters* **11**, 269–271 (1967).
35. N. F. Foster, "Ultrahigh frequency cadmium sulfide transducers," *IEEE Trans. on Sonics and Ultrasonics* **SU-11**, 63–68 (1964).
36. J. de Klerk and E. F. Kelley, "Coherent phonon generation in the gigacycle range via insulating cadmium sulfide films," *Appl. Phys. Letters* **5**, 2–3 (1964).
36a. T. R. Sliker and D. A. Roberts, "A thin film CdS–quartz composite resonator," *J. Appl. Phys.* **38**, 2350–2358 (1967).
37. D. L. White, "Depletion layer transducer—A new high frequency ultrasonic transducer," 1961 *IRE Internat'l Conv. Rec.* pt. 6, vol. 9 (1961), pp. 304–309.
38. N. F. Foster, "Diffusion layer ultrasonic transducer," *J. Appl. Phys.* **34**, 990–991 (1963).

39. A. R. Hutson, J. H. McFee, and D. L. White, "Ultrasonic amplification in CdS," *Phys. Rev. Letters* **7**, 237–239 (1961).
40. S. B. Austerman, D. Berlincourt, and H. H. A. Krueger, "Polar properties of BeO single crystals," *J. Appl. Phys.* **34**, 339–341 (1963).
41. D. F. Crisler, J. J. Cupal, and A. R. Moore, "Dielectric, piezoelectric, and electromechanical coupling constants of zinc oxide crystals," *Proc. IEEE* **56**, 225–226 (1968).
42. T. B. Bateman, "Elastic moduli of single-crystal zinc oxide," *J. Appl. Phys.* **33**, 3309–3312 (1962).
43. D. A. Berlincourt, H. Jaffe, and L. R. Shiozawa, "Electroelastic properties of the sulfides, selenides, and tellurides of zinc and cadmium," *Phys. Rev.* **129**, 1009–1017 (1963).
44. A. R. Hutson, "Piezoelectric devices utilizing aluminum nitride," U.S. Patent 3,090, 876, May 21, 1963.
45. A. W. Warner, G. A. Coquin, A. H. Meitzler, and J. L. Fink, "Piezoelectric Properties of $Ba_2NaNb_5O_{15}$," *Appl. Phys. Letters* **14**, 34–35 (1969).
46. A. W. Warner, "New piezoelectric materials," presented at the 19th Frequency Control Symp., sponsored by U.S. Army Electronics Command, Fort Monmouth, N. J., April, 1965.
47. M. Marezio, "The crystal structure of $LiGaO_2$," *Acta Cryst.* (*Internat.*) **18**, 481–484 (1965).
48. E. J. Charlson and G. Mott, "Dynamic measurement of the piezoelectric and elastic constants of gallium arsenide," *Proc. IEEE* **51**, 1239 (1963).
49. T. B. Bateman, H. J. McSkimin, and J. M. Whelan, "Elastic moduli of single-crystal gallium arsenide," *J. Appl. Phys.* **30**, 544–545 (1959).
50. M. Onoe, A. W. Warner, and A. A. Ballman, "Elastic and piezoelectric characteristics of bismuth germanium oxide $Bi_{12}GeO_{20}$," *IEEE Trans. on Sonics and Ultrasonics* **SU-14**, 165–167 (1967).
51. P. Egli, "A survey of inorganic piezoelectric materials," *American Minerologist* **33**, 622–633 (1948).
52. A. W. Warner, M. Onoe, and G. A. Coquin, "Determination of elastic and piezoelectric constants for crystals in class (3m)," *J. Acoust. Soc. Am.* **42**, 1223–1231 (1967).
53. A. W. Warner and A. A. Ballman, "Low temperature coefficient of frequency in a lithium tantalate resonator," *Proc. IEEE* **55**, 450 (1967).
54. T. R. Sliker and D. J. Koneval, "Frequency–temperature behavior of X-cut lithium tantalate resonators, *Proc. IEEE* **56**, 1402 (1968).
55. E. Fatuzzo, G. Harbeke, W. J. Merz, R. Nitsche, H. Roelschi, and W. Ruppel, "Ferroelectricity in SbSI," *Phys. Rev.* **127**, 2036–2037 (1962).
56. R. Nitsche, H. Roelschi, and P. Weld, "New ferroelectric V, VI, and VII compounds of the SbSI type," *Appl. Phys. Letters* **4**, 210–211 (1964).
57. D. A. Berlincourt, H. Jaffe, W. J. Merz, and R. Nitsche, "Piezoelectric effect in the ferroelectric range in SbSI," *Appl. Phys. Letters* **4**, 61–63 (1964).
58. G. Quentin and J. M. Thuillier, "Piezoelectric properties of tellurium by electromechanical resonance," *Solid State Commun.* **2**, 115–117 (1964) (in French).
59. H. Gobrecht, H. Hamisch, and A. Tausend, "The piezoelectric effect in selenium," *Z. Phys.* (*Germany*) **148**, 209–217 (1957).
60. B. Jaffe, R. S. Roth, and S. Marzullo, "Piezoelectric properties of lead zirconate–lead titanate solid-solution ceramics," *J. Appl. Phys.* **25**, 809–810 (1954).

61. D. A. Berlincourt, C. Cmolik, and H. Jaffe, "Piezoelectric properties of polycrystalline lead titanate zirconate compositions," *Proc. IRE* **48**, 220–229 (1960).
62. D. Schofield and R. F. Brown, "An investigation of some barium titanate compositions for transducer applications," *Canad. J. Phys.* **35**, 594–607 (1957).
63. C. S. Brown, R. C. Kell, R. Taylor, and L. A. Thomas, "Piezoelectric materials," *Proc. Instn Elect. Engrs (GB)* **109B**, 99–114 (1962).
64. R. E. Jaeger and L. Egerton, "Hot pressing of potassium–sodium niobates, *J. Am. Ceram. Soc.* **45**, 209–213 (1962).
65. F. Kulcsar, "Electromechanical properties of lead titanate zirconate ceramics with lead partially replaced by calcium or strontium," *J. Am. Ceram. Soc.* **42**, 49–51 (1959).
66. T. Ikeda, "Studies on (Ba, Pb)(Ti, Zr)O_3 system," *J. Phys. Soc. Japan*, 168–174 (1959).
67. B. Jaffe, R. S. Roth and S. Marzullo, "Properties of piezoelectric ceramics in the solid-solution series lead titanate–lead zirconate-lead oxide: tin oxide and lead titanate–lead hafnate," *J. Res. Nat. Bur. Stand.* **55**, 239–254 (1955).
68. F. Kulcsar, "Electromechanical properties of lead titanate zirconate ceramics modified with certain three- or five-valent additions," *J. Am. Ceram. Soc.* **42**, 343–349 (1959).
69. H. Jaffe, "Properties of ferroelectric ceramics in the lead titanate zirconate system," *Proc. Inst. Elect. Eng. (GB)* **109B**, Suppl. 22, 351–354 (1961).
70. R. Gerson and H. Jaffe, "Electrical conductivity in lead titanate zirconate ceramics," *J. Phys. Chem. Solids* **24**, 979–984 (1963).
71. D. Berlincourt, "Behavior of piezoelectric ceramics under various environmental and operation conditions of radiating sonar transducers," *U.S. Navy J. of Underwater Acoust.* **15**, 266–283 (1965).
72. H. H. A. Krueger, "Stress sensitivity of piezoelectric ceramics: Pt. II. Heat treatment," *J. Acoust. Soc. Am.* **43**, 576–582 (1968).
73. H. H. A. Krueger, "Stress sensitivity of piezoelectric ceramics: Pt. I. Sensitivity to compressive stress parallel to the polar axis, *J. Acoust. Soc. Am.* **42**, 636–645 (1967).
74. D. Schofield and R. F. Brown, "An investigation of some barium titanate compositions for transducer applications," *Can. J. Phys.* **35**, 594–607 (1957).
75. R. F. Brown, "Effect of two-dimensional mechanical stress on the dielectric properties of ceramic barium titanate and lead zirconate titanate," *Can. J. Phys.* **39**, 741–753 (1961).
76. R. F. Brown and G. W. McMahon, "Material constants of ferroelectric ceramics at high pressure," *Can. J. Phys.* **40**, 672–674 (1962).
77. R. Y. Nishi and R. F. Brown, "Behavior of piezoceramic projector materials under hydrostatic pressure," *J. Acoust. Soc. Am.* **36**, 1292–1296 (1964).
78. R. Y. Nishi, "Effects of one-dimensional pressure on the properties of several transducer ceramics," *J. Acoust. Soc. Am.* **40**, 486–495 (1966).
79. D. Berlincourt and H. H. A. Krueger, "Domain processes in lead titanate zirconate and barium titanate ceramics," *J. Appl. Phys.* **30**, 1804–1810 (1959).
80. H. H. A. Krueger, "Stress sensitivity of piezoelectric ceramics: Pt. III. Sensitivity to compressive stress perpendicular to the polar axis," *J. Acoust. Soc. Am.* **43**, 583–591 (1968).
81. A. W. Warner, J. G. Bergman, Jr., D. A. Pinnow, and G. R. Crane, "Piezoelectric and photoelastic properties of lithium iodate," *J. Acoust. Soc. Am.* **47**, 791–794 (1970).
82. S. Haussühl, "Piezoelectric and electric behavior of lithium iodate" (in German), *Phys. Stat. Sol.* **29**, K159–161 (1968).

Chapter 3

PIEZOELECTRIC TRANSDUCER MATERIALS AND TECHNIQUES FOR ULTRASONIC DEVICES OPERATING ABOVE 100 MHz

A. H. Meitzler

Bell Telephone Laboratories
Murray Hill, New Jersey

1. INTRODUCTION

1.1. Scope of Chapter

In the field of ultrasonics, the frequency range above 100 MHz was until a few years ago largely the province of research scientists concerned with studies of acoustic losses in solids. However, this frequency range has since become important for engineers concerned with certain ultrasonic devices such as dispersive and nondispersive ultrasonic delay lines, ultrasonic light deflectors, and ultrasonic light modulation devices. There are already in existence practical ultrasonic delay lines(1) for the storage of digital data signals at bit rates as high as 100 Mbits/sec. Similarly, light modulators(2) and light deflectors(3,4) operating at frequencies of several hundred MHz have been built and light modulators, light deflectors, and delay lines requiring total 3 dB bandwidths of several hundred MHz are within the reach of existing technological capabilities.

Figure 1 shows three examples of high-frequency ultrasonic and acoustooptical device structures with transducer configurations of the sort that are of primary interest for this chapter. The two rectangular bars are typical of structures used for high-frequency delay lines and light deflectors; the bar in the center has a bonded, thin-plate, single-crystal transducer of

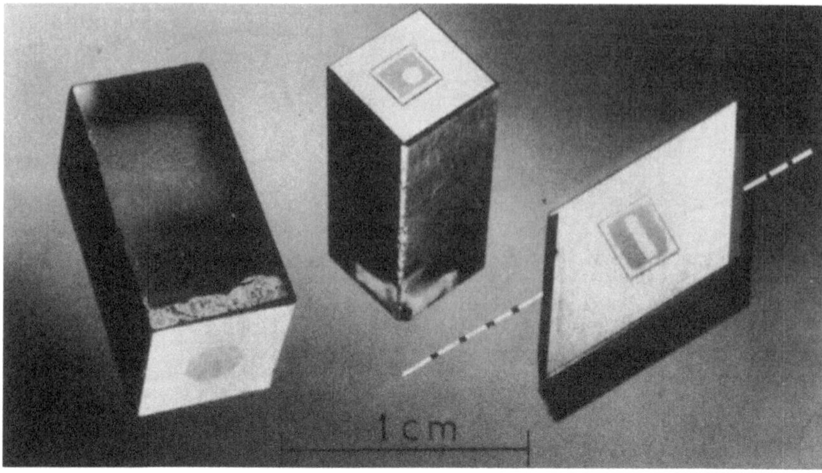

Fig. 1. High-frequency ultrasonic and acoustooptic device structures using sputtered film and bonded plate transducers.

lithium niobate ($LiNbO_3$), the bar on the left has a sputtered-film zinc oxide (ZnO) transducer. The structure on the right has parallel end faces inclined at an angle to the length axis of the bar (indicated by the dashed line) in order to permit a polarized light beam to enter and exit at the Brewster angle. This device structure is used as a light modulator in a mode-locked laser cavity. The transducer is again $LiNbO_3$. All three examples shown use fused quartz as the acoustic medium, although at frequencies above 100 MHz, a number of single crystal materials are presently in use in ultrasonic and acoustooptical applications.

The major objective of the present chapter will be to review the piezoelectric materials that are useful in high-frequency transducers. In addition, there are a variety of experimental techniques that have been developed especially for use in the fabrication and evaluation of transducers for high-frequency ultrasonic devices. An attempt will be made to describe the more important techniques and to provide at least a brief mention of the techniques not covered in detail and an indication of appropriate sources of additional information.

1.2. Results from the Equivalent Circuit Analysis of High-Frequency Ultrasonic Devices Using Piezoelectric Transducers

The starting point for the discussion of transducer materials for high-frequency ultrasonic devices is the basic physical configuration of the transducer structure and the associated equivalent electrical circuit.[5-14] Figure 2a shows the basic arrangement for a high-frequency piezoelectric transducer of the type commonly used in device structures. The transducer is a thin piezoelectric layer of thickness t, in contact on one face with a transmission medium into which it radiates acoustic energy. In order to obtain maximum driving field in the piezoelectric layer, metal electrode layers are used in immediate contact with the piezoelectric layer.

The acoustic properties of the transmission medium are given completely for present purposes by its specific mechanical impedance Z_m, which is simply the product of the density and the appropriate elastic wave velocity, $Z_m = \varrho_m V_m$. The acoustic properties of the transducer layer are specified by similar quantities, $Z_0 = \varrho_0 V_0$. In addition, the equivalent circuit analysis indicates two additional quantities are needed: k, the piezoelectric coupling factor; and K, the relative dielectric constant which involves, specifically, the ratio of the constant-strain permittivity ε^S of the piezoelectric medium and the free-space dielectric permittivity ε_0, ($K = \varepsilon^S/\varepsilon_0$).

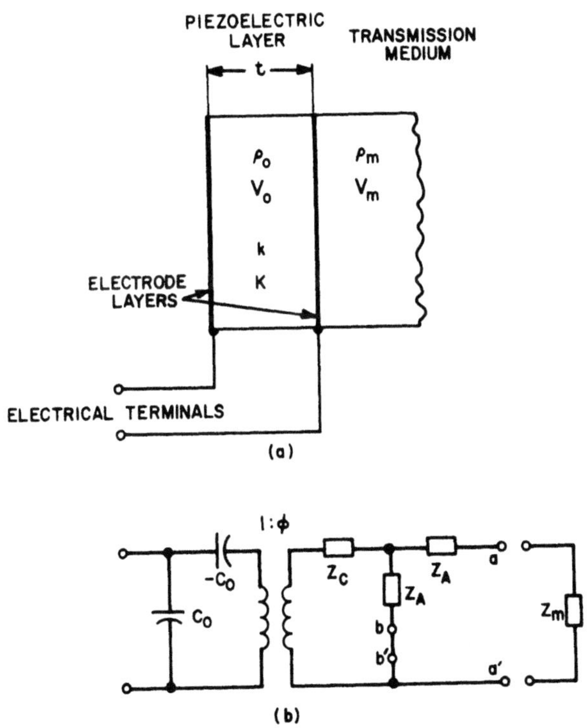

Fig. 2. (a) Basic configuration of a layer transducer and an acoustic medium. (b) The equivalent electrical circuit.

With the basic parameters of the transducer layer and the transmission medium introduced, it is possible to define the elements of the equivalent circuit shown in Fig. 2b. The quantity C_0 is the shunt capacitance of the transducer and is given by

$$C_0 = \varepsilon^S \cdot \frac{A}{t} = K\varepsilon_0 \cdot \frac{A}{t} \tag{1}$$

where A is the area of the active transducer, usually determined by the dimensions of the outer electrode layer, and t is the thickness of the piezoelectric layer. The impedances Z_A and Z_C are given by

$$Z_A = (jZ_0/2) \tan \gamma \tag{2}$$

and

$$Z_C = jZ_0/\sin \gamma \tag{3}$$

in which $\gamma = \pi\omega/\omega_0$; and the transformer ratio φ is given by

$$\varphi^2 = \omega_0 C_0 Z_0 k^2/\pi \tag{4}$$

The quantity ω_0 is $2\pi f_0$, f_0 being the frequency at which the transducer thickness is one-half wavelength. The equivalent circuit of the transducer has two pairs of mechanical terminals a–a' and b–b'. The pair b–b' is shown as short-circuited since in Fig. 1a only one major surface of the transducer is loaded.

The equivalent circuit of Fig. 1b, due originally to W. P. Mason,[5] has been found by many authors[5–16] to be an excellent representation with which to interpret many experimental observations provided that the transducer operates under conditions producing a single pure mode of vibration. This is usually the case when (1) the driving field is applied in the thickness direction by electrode layers immediately adjacent to the piezoelectric layer, (2) the lateral dimensions of the transducer are much larger than the thickness dimension, and (3) the orientation of the material in the piezoelectric layer is deliberately controlled or selected so that an electric field in the thickness direction couples to a pure thickness-shear or thickness-dilatational mode of vibration.

A single transducer by itself has, of course, one electrical port and can be characterized electrically by its impedance or admittance as a function of frequency. With the three basic transducer parameters (K, k, Z_0) and the mechanical impedance of the medium, it is possible to calculate from the equivalent circuit the normalized input conductance and susceptance as

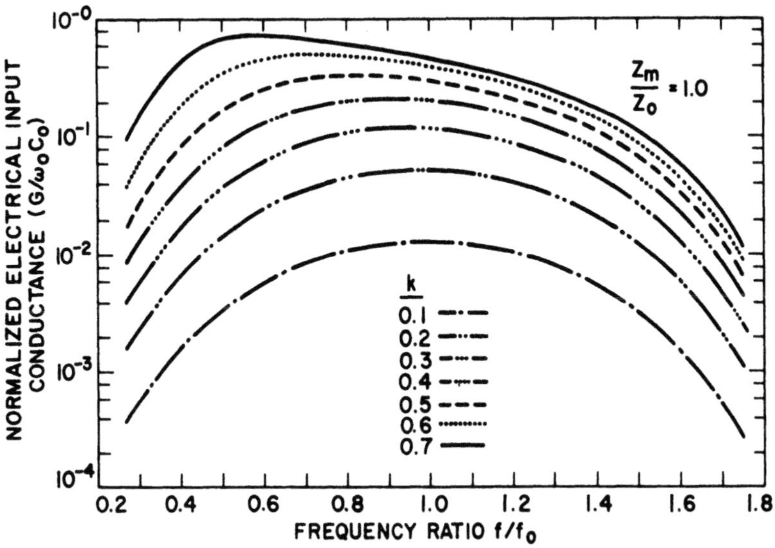

Fig. 3. Normalized electrical input conductance *vs* frequency ratio with coupling factor as a parameter (after Meitzler and Sittig[13]).

functions of frequency for any combination of materials of interest. It is often desirable to operate in the range Z_m/Z_0 approximately equal to 1. Figure 3 containes a set of curves showing normalized conductance $G/\omega_0 C_0$ as a function of normalized frequency f/f_0 with k varying in steps of 0.1 from 0.1 to 0.7 and with $Z_m/Z_0 = 1$. For some applications, for example, one requiring specific bandpass characteristics,[15] operation at $Z_m/Z_0 \neq 1$ is indicated. The conductance curves show that for high coupling factors ($k \approx 0.7$) the normalized conductance ratio near $f/f_0 = 1$ is close to 0.5. For low coupling factors ($k = 0.1$) the ratio is approximately 10^{-2}.

The explicit equations for the conductance or the impedance seen at the electrical input terminals as functions of frequency are complicated; but for the special case $f = f_0$, the impedance equation simplifies[14] to

$$Z_i = \frac{1}{\omega_0 C_0}\left(-j + \frac{4k^2 Z_0}{\pi Z_m}\right) \tag{5}$$

which indicates that the impedance for a bonded transducer at this frequency is equivalent to a series connection of the capacitance C_0 and a resistance

$$R_i = 4k^2 Z_0/\pi Z_m \omega_0 C_0 \tag{6}$$

Usually, the variation of insertion loss (or gain) with frequency is a

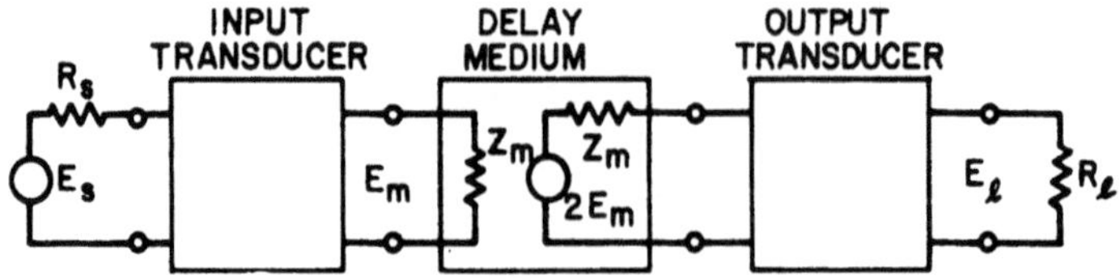

Fig. 4. Equivalent circuit of an ultrasonic delay line with separate input and output transducers.

quantity of more immediate interest than conductance or susceptance. Insertion loss is defined for a two-port device. Ultrasonic devices frequently involve two separate transducers so that an analysis of a device involving two transductions can be carried out by means of the equivalent circuit of Fig. 4. This circuit is strictly correct only for a lossless delay medium having a delay length long enough, compared to the duration of the input signal, that reflections of acoustic waves can be neglected. With the symbols of Fig. 4, the conventional definition of insertion loss (IL) for the network becomes:

$$\text{IL (in dB)} = 20 \log_{10} \frac{E_s}{E_l} + 20 \log_{10} \frac{R_s}{R_s + R_l} \tag{7}$$

The insertion gain is merely the negative of the insertion loss.

Insertion loss is of practical significance because it is more easily measured with precision than the input conductance and, if measured with fixed resistive or reactive terminations, characterizes the device under actual operating conditions.

Although insertion loss is defined for a two-port device, insertion loss measurements can be performed meaningfully on ultrasonic device structures with a single transducer. If the transducer is on one face of a medium with flat parallel end faces and measured under appropriate conditions by an rf pulse–echo technique,[14] a loss measurement directly identifiable with the insertion loss of a symmetrical two-port device connected between resistive terminations can be obtained. This "single-ended" measuring technique is frequently used in evaluating high-frequency ultrasonic transducers.

Figure 5 shows a family of curves for insertion gain plotted as a function of normalized frequency for two identical piezoelectric transducers bonded to a medium with impedance ratio $Z_m/Z_0 = 1$. The termination conditions used were $R_s = R_l = (\omega_0 C_0)^{-1}$, which corresponds to the resistive termination conditions that provide maximum power transfer at f_0. (Ideal piezoelectric transducers, loaded on only a single face, as indicated in Fig. 2,

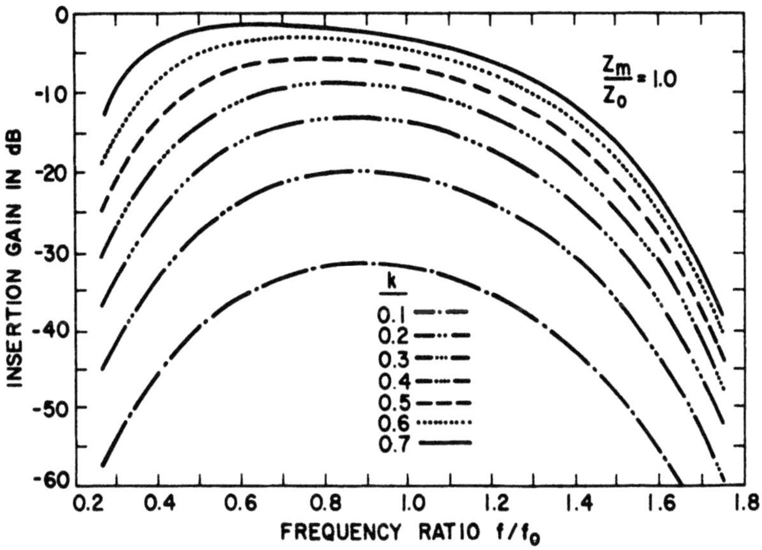

Fig. 5. Insertion gain in dB *vs* frequency ratio with coupling factor as a parameter (after Meitzler and Sittig[13]).

can provide zero insertion loss regardless of k or Z_m/Z_0 at any frequency for which they can be matched to the source and load.) The fixed resistive termination conditions chosen are also of interest because they correspond to wide-band (low electrical Q) conditions and serve admirably to illustrate the general point, that for $Z_m/Z_0 \leq 1$, the maximum level and bandwidth of the transmission increases as the k of the transducer increases. This observation is in accord with what one might intuitively expect; however, complications in loss characteristics do arise for $Z_m/Z_0 > 1$ that limit the validity of this generalization.[14] By virtue of the happenstance that most transducer materials have Z_0's larger than the Z_m's of glasses and other commonly used ultrasonic media, the generalization is, however, usually correct.

The dependence of the insertion loss of an ultrasonic device upon the material constants of the transducer material and the acoustic medium can be further demonstrated by the following approximate formula

$$\text{IL (in dB)} \simeq 20 \log_{10}\left(\frac{\pi}{8k^2}\,\frac{Z_m}{Z_0}\right) \tag{8}$$

which applies at the frequency $\omega = \omega_0$ for an ultrasonic device with two ideal unbacked piezoelectric transducers operated under resistive symmetrical termination conditions corresponding to $R_s = R_l = (\omega_0 C_0)^{-1}$.

Further, this relation is restricted to transducers with low coupling factors ($k < 0.3$). In addition to the loss associated with electrical termination conditions and capable of being estimated by Eq. (8), a real ultrasonic device has contributions to IL from attenuation in the acoustic medium and from diffraction effects associated with the finite lateral dimensions of the transducer. For the case of a device using only one transducer, such as an ultrasonic light deflector, the insertion loss of the transducer itself is one-half the value estimated by Eq. (8).

Variation of insertion loss as a function of frequency is emphasized in the existing literature rather than the phase characteristic of a structure incorporating piezoelectric transducers. Nevertheless the dependence of a phase upon k and Z_m/Z_0 for an ultrasonic device can be important as, for example, in the case of digital wide-band delay lines, which often require a linear phase response. Again, the equivalent circuit analysis forms a convenient and reliable basis for evaluating in detail the influence of different combinations of material properties on device performance characteristics. Detailed calculations of this sort applied to the design of ultrasonic delay lines can be found in a paper by Sittig.[15]

1.3. Properties of Transducer Materials and Acoustic Materials of Interest for High-Frequency Applications

Based on preceding observations it is possible to identify desirable attributes of piezoelectric materials for high-frequency ultrasonic devices. Generally speaking, these materials should combine a high coupling factor and low dielectric constant with a value of mechanical impedance approximately equal to common acoustic and acoustooptical materials. Berlincourt[17] has defined a figure of merit given by

$$M = kN/\varepsilon^S \tag{9}$$

but as the discussion of the preceding subsection has indicated, the ratio of the transducer and acoustic medium mechanical impedances as well as K, k, and Z_0, enters into the picture and hence a figure of merit involving only the parameters of the transducer materials has limited usefulness. In line with this point of view, two principal survey tables are presented in this chapter, one summarizing the properties of important transducer materials presently available (Table II) and the other summarizing the properties of commonly used acoustic media for high-frequency devices (Table III). In the analysis of the performance of ultrasonic devices, the clearest understanding is obtained by using the equivalent circuit with the required parameters to calculate whatever quantities are of interest. With modern computer facilities, this approach is much more practical than it was when the equivalent circuit formulation was originally derived.

1.3.1. A Survey of Transducer Materials Particularly Suitable for High-Frequency Devices

High-frequency ultrasonic devices have benefited from the development of new single crystal and ceramic transducer materials and undoubtedly will continue to do so. The changing nature of the materials available can be illustrated strikingly by citing the materials available and competitive with one another for device applications in 1969 with materials that were available ten years earlier. Table I provides such a listing, with materials classified into three categories.

Table I. Materials Useful as Piezoelectric Transducers in High-Frequency Devices

	ca. 1959	*ca.* 1969
Piezoelectric materials		
Single crystals		
a. naturally occurring	SiO_2	SiO_2
b. synthetically grown		$LiGaO_2$
		SiO_2
		$Bi_{12}GeO_{20}$
Compound semiconductor films		
Evaporated—polycrystalline with preferential orientation or single crystal, epitaxially grown		CdS
Sputtered—polycrystalline with preferential orientation		ZnO
Ferroelectric materials		
Poled ceramics	MBT[a]	SPN[c]
	PZT-5[b]	PZT-7A[b]
Single crystals		$LiNbO_3$
		$LiTaO_3$
		$Ba_2NaNb_5O_{15}$

[a] A modified $BaTiO_3$ ceramic, developed at Bell Telephone Laboratories, Inc. and consisting of 80 wt. % $BaTiO_3$, 12 wt. % $PbTiO_3$ and 8 wt. % $CaTiO_3$.
[b] Trademark, Clevite Corporation.
[c] Ceramic $Na_{0.5}K_{0.5}NbO_3$, developed at Bell Telephone Laboratoires, Inc.

It is immediately obvious that over the decade from 1959 to 1969 not only the number of materials changed significantly but also the types of materials changed. In 1959, the choice of transducer materials for high-frequency ultrasonic devices such as delay lines was principally between crystalline quartz and poled ferroelectric ceramic materials. By 1969, the naturally occurring piezoelectric single-crystal materials were augmented by synthetically grown piezoelectric and ferroelectric materials and by substantially improved poled ferroelectric ceramics. In addition, thin-film transducer techniques, using principally cadmium sulfide (CdS) and zinc oxide (ZnO), were developed and became the most useful means of achieving efficient transducers at frequencies above 500 MHz.

The transducer materials placed in the two columns (one headed 1959 and the other 1969) do not, of course, represent all of the materials available at either time. They are selected as being those most worthy of serious consideration in applications appropriate to the time, with each offering certain advantages and disadvantages. Many of these advantages and disadvantages will be made clear in later sections of this chapter.

Table II lists all the piezoelectric and ferroelectric materials of practical interest for high-frequency transducers for which a substantial amount of information is known. Their number is considerably smaller than the large number of crystalline and ceramic materials reviewed in the chapter by D. Berlincourt contained in this volume. The basis of selection is discussed in Section 1.3.2.

The information presented in Table II summarizes all the parameters required in an equivalent circuit analysis of the performance of a given material as a high-frequency thickness-mode transducer. The first column designates the material by its nominal chemical formula, except in the case of the ceramic compounds. The point-group designation of the material, in Hermann–Mauguin symbols, is presented in parentheses. The symbols [D] or [S] indicate whether the information on a given line applies to thickness-dilatational or thickness-shear transducers. The second column indicates the orientation required for a plate of the material to operate in the mode indicated. Several different types of designations are used, but the ones that are shown are fairly standard and are explained in detail in the text relating to the particular material. The third column presents the "effective" value of the electromechanical coupling factor, k used in this mode. In the context of this chapter, the value of k required in Eq. (4) to predict correctly transducer performance from the equivalent circuit becomes, in a sense, an operational definition of k. When k is related to specific dimensionless combinations of fundamental material constants,

Table II. Materials for High-Frequency Piezoelectric Transducers (ca. 1969)

Material (point group)	[Mode]	Orientation	Coupling factor, k	Relative dielectric constant, K	Frequency constant $f_0 \cdot t$, MHz · μm	Mechanical impedance Z_0, 10^6 kg/s · m²	Ref.
Piezoelectric materials							
Single crystal							
SiO_2 (32)	[*D*]	X	0.098	4.58	2850	15.1	10
	[*S*]	Y	0.137	4.58	1925	10.2	
	[*S*]	AC	0.083	4.58	1650	8.8	
$LiGaO_2$ (*mm*2)	[*D*]	Z	0.30	8.5	3130	26.2	51
	[*S*]	X	0.17	7.0	2550	21.4	
	[*S*]	Y	0.17	6.0	2900	24.2	
$Bi_{12}GeO_{20}$ (23)	[*D*]	(111)	0.155	38	1670	30.6	53
	[*S*]	(110)	0.235	38	850	15.6	
$LiIO_3$ (6)	[*D*]	0°	0.51	6	2066	18.5	55, 56
	[*S*]	90°	0.60	8	1260	11.3	
Films							
CdS (6*mm*)	[*D*]	0°	0.154	9.53	2750	21.7	16
	[*S*]	90°	0.188	9.02	900	8.69	
	[*S*]	39.7°	0.212	9.33	1050	10.2	
ZnO (6*mm*)	[*D*]	0°	0.282	8.84	3200	36.4	16, 37
	[*S*]	90°	0.259	8.33	2880	16.4	
	[*S*]	43.0°	0.322	8.63	1620	18.4	
AlN (6*mm*)	[*D*]	0°	0.20	8.5	5200	34.0	82

Table II (*continued*)

Material (point group)	[Mode]	Orientation	Coupling factor, k	Relative dielectric constant, K	Frequency constant $f_0 \cdot t$, MHz · μm	Mechanical impedance Z_0, 10^6 kg/s · m²	Ref.
Ferroelectric materials							
Poled ceramics							
MBT (6*mm*)	[*D*]	0°	0.30	395	2880	31.3	96, 97
	[*S*]	90°	0.30	365	1700	18.4	
PZT-5 (6*mm*)	[*D*]	0°	0.49	635	2170	33.4	96, 97
	[*S*]	90°	0.69	730	1120	17.3	
SPN (6*mm*)	[*D*]	0°	0.46	310	3470	31.3	17, 70
	[*S*]	90°	0.65	545	1880	16.9	71
PZT-7A (6*mm*)	[*D*]	0°	0.50	235	2400	33.8	17, 96
	[*S*]	90°	0.67	460	1250	17.6	97
Single Crystals							
$LiNbO_3$ (3*m*)	[*D*]	Z	0.17	29	3660	34.4	61
	[*D*]	35°Y	0.49	39	3700	34.8	
	[*S*]	163°Y	0.62	43	2280	21.4	
	[*S*]	X	0.68	44	2400	22.3	
$LiTaO_3$ (3*m*)	[*D*]	Z	0.19	43	3040	45.3	61
	[*D*]	47°Y	0.29	42	3700	55.1	
	[*S*]	165°Y	0.41	41	2280	34.0	
	[*S*]	X	0.44	41	2100	31.3	
$Ba_2NaNb_5O_{15}$ (*mm*2)	[*D*]	Z	0.57	32	3075	32.6	68, 69
	[*S*]	X	0.21	222	1820	19.3	
	[*S*]	Y	0.25	227	1830	19.4	

it is written with various letter or number subscripts. For example, k_t is the symbol used often in connection with thickness-extensional modes and k_{15} often in connection with thickness-shear modes. The standard subscripted symbols for k apply only to axial-plate cuts and, since many of the plates of interest are rotated cuts, no attempt was made to retain subscripts in connection with the values of k listed in the table. The reader interested in more details related to definitions of k is referred to the chapter by D. Berlincourt.

A similar situation exists with respect to the value of K presented in the fourth column. In this case, Eq. (1) becomes effectively an operational definition. The fifth column presents the value of the frequency constant for what is conventionally termed the *parallel resonance frequency*, f_0. This is the frequency at which the transducer is exactly an acoustic half-wavelength thick. Note that, since this is the case,

$$f_0 \cdot t = f_0(\lambda_0/2) = V_0/2 \tag{10}$$

The sixth column is the specific mechanical impedance of the transducer material, defined earlier as $Z_0 = \varrho_0 V_0$; hence the value of ϱ_0, if desired explicitly, can be extracted from the information in the last two columns

$$\varrho_0 = Z_0/2f_0 \cdot t \tag{11}$$

The numbers in the last column refer to the references listed at the end of this chapter.

In addition to the elastic, dielectric, and piezoelectric properties, a real transducer material has other properties which are not easily specified by numbers. There are, for example, the considerations of availability, workability or ease of fabrication, and susceptibility to damage both in processing and in final application. In some cases, temperature dependence of material constants and their influence on the performance of the finished transducer become important. Finally, there is the consideration of cost which, for the designer of practical devices, is of great significance. Some of these points will be discussed when the individual transducer materials and techniques are described in the following subsections.

1.3.2. Discussion of the Basis of Selection for the Inclusion of Materials in the Survey Table of Transducer Materials

For practical high-frequency device applications the most useful transducer techniques have involved bonded thin plates or evaporated and sputtered thin films. The bonded thin-plate techniques are useful up to approximately 500 MHz[18,19] and the thin-film transducer techniques are useful for frequencies considerably beyond one GHz.[16,20–22] Presently (1970) thin-film transducer techniques have achieved maximum coupling factors of approximately 0.25 for dilatational-mode transducers of ZnO, while the thin-plate transducers, using the recently developed ferroelectric material $LiNbO_3$, have achieved substantially the bulk values, 0.49 for thickness-dilatational and 0.68 for thickness-shear.

The materials ZnO and CdS are members of a large group of piezoelectric semiconductor compounds which have become of interest in recent years not only as transducer materials but also as materials useful in obtaining ultrasonic amplification.[23,24] In addition, piezoelectric semiconductor compounds have been used to obtain two novel types of transducer structures; the depletion layer transducer and the diffusion-layer transducer, both originally conceived by D. L. White.[25,26] Diffusion-layer transducers were developed to a significant level of performance by N. F. Foster.[27] The early work with depletion-layer and diffusion-layer transducers stimulated the development of techniques for realizing transducer action from deposited films of semiconductor materials.[20,22] While conceptually important, both the depletion-layer and the diffusion-layer transducer techniques no longer appear to be competitive with thin-plate and thin-film techniques for most device structures, although it should be noted that diffusion-layer transducers have had some use in high-frequency delay lines and in experimental ultrasonic amplifiers with integrated transducer structures[28–32] and may find wider future applications with improvements in materials and techniques. The principal reasons for the lack of application of these semiconductor transducer methods appear to be related to problems of controlling both the initial properties of the materials and the long-term electrical characteristics of the completed transducers. Along with these problems, semiconductor transducers often involve losses due to conductivity mechanisms and, at best, present materials are limited to moderate values of coupling factors.

Although a number of the semiconductor compounds can be grown in the form of large single crystals from which transducer plates can be cut, these materials have not been found to be useful in thin-plate transducers for device applications. In crystalline form, these materials are relatively soft, fragile, and difficult to work and usually, because no large-scale applications exist, are obtainable only in limited quantities on an experimentally-grown basis. Consequently, single-crystal semiconductor materials have suffered from variability in material parameters. For example, one of the most potentially attractive materials in the group of semiconductor compounds is ZnO. A number of different research groups have worked with this material and have reported difficulties such as barrier-layer effects and effects due to c-axis polarity reversals.[33–37] Thus the present state of affairs is such that, even though ZnO is quite successful as a film-transducer material, the properties of the single-crystal are so poorly controlled that values of the electroelastic constants appearing in the literature over the last few years have shown substantial variation.

The semiconductor compounds are most useful as transducers when used as thin films. Here the relatively simple structures of the basic crystalline solids and, in the case of some materials like CdS, low decomposition temperatures combine to make these materials the most attractive for the formation of high-frequency transducers by evaporating or sputtering techniques. The problem of conductivity loss by Joule heating is still frequently present, but by proper techniques can be controlled; in the case of CdS and ZnO, films with negligible conductivity loss can be achieved. The polycrystalline structure of the evaporated and sputtered films undoubtedly contribute to the reduction of undesirable effects that depend upon mobilities of conduction electrons. The polycrystalline structure of these films is, at the same time, accompanied by the disadvantage that, although a high degree of c-axis orientation can be obtained, some randomness in orientation and polarity does remain and, as a practical consequence, films formed by these techniques often show values of coupling factor 10 to 20% less than the single-crystal values.

The discussion of material properties at this point has been from the point of view of explaining the basis on which the entries of Tables I and II were selected. A more detailed discussion of the properties of these materials relevant to high-frequency transducer applications will be given in later subsections.

1.3.3. A Survey of Acoustic Properties for Media Commonly Used in High-Frequency Ultrasonic and Acoustooptic Devices

High-frequency ultrasonic and acoustooptic devices necessarily involve more than just transducer materials. There are, in addition, the amorphous or crystalline solids used to carry the mechanical wave motions and there are the films used as the electrode and bonding layers. The bonding layers may be plastics (such as epoxy resins), glasses, (such as low melting point arsenic–selenium–sulfur compounds[38]), or metals. The electrode layers are invariably metals. Usually the thin metallic layers used for electrode or bonding layer are polycrystalline and, hence, behave as simple isotropic solids.

Table III. Properties of Acoustic Media Used in High-Frequency Devices

Material (direction)	Velocities, m/sec			Impedances, 10^6 kgm/s · m²			Attenuation, (dB/cm) at [MHz]		Ref.
	V_D	V_{S_1}	V_{S_2}	Z_D	Z_{S_1}	Z_{S_2}	α_D	α_S	
Amorphous solids									
Glasses									
Fused quartz	5960	3760	—	13.1	8.29	—	(0.2) [100] (3.5) [500]	(0.3) [100] (5.0) [500]	98, 99
T-40	4310	2580	—	14.6	8.7	—	—	(3.0) [100]	100, 101
As–S–Se glass	—	0.90	—	—	2.32	—	—	—	38
Electrode and Bonding Layers									
Ag	3650	1610	—	38.0	16.7	—	—	—	102
Al	6420	3040	—	17.3	8.2	—	—	—	102
Au	3240	1200	—	62.5	23.2	—	—	—	102
Cu	5010	2110	—	40.6	18.3	—	—	—	102
In	2300	1440	—	17.0	10.5	—	—	—	10
Epoxy	2600	1220	—	2.86	1.34	—	—	—	103
Phenyl benzoate	—	—	—	3.38	0.48	—	—	—	107
Crystalline solids									
Al_2O_3 [10.0]	11030	6780	5720	43.8	26.9	22.7	(0.1) [500]	—	104
$LiNbO_3$ [10.0]	6550	4760	4030	30.8	22.4	18.9	(0.2) [500]	—	57
$LiTaO_3$ [10.0]	5550	4210	3370	41.4	31.2	35.1	(0.1) [500]	—	57
MgO [100]	9230	6600	6600	33.0	23.6	23.6	—	—	104
$PbMoO_4$ [001]	3750	1990	1990	26.0	13.8	13.8	(1.0) [500]	—	106
SiO_2 [10.0]	5730	5160	3310	17.2	13.7	8.8	(1.5) [500]	—	104
YAG [001]	8560	5030	5030	39.0	22.9	22.9	(0.05) [500]	—	105
YIG [001]	7210	3840	3840	37.2	19.8	19.8	(0.05) [500]	—	105

Table III is provided as a brief listing of the acoustic properties of some common materials, other than transducer materials, used in ultrasonic and acoustooptic devices. The isotropic solids have two velocities, the dilatational, V_D, and the shear, V_{S_1}, respectively. Crystalline elastic solids are much more complicated with three independent modes of stress wave propagation possible along any arbitrary direction.[39] The propagation direction is listed for a crystalline solid in square brackets and the wave velocities, one for the dilatational mode and two for the shear modes with velocities V_{S_1} and V_{S_2}, are listed in the next three columns of the table. The corresponding specific mechanical impedances are listed following the velocities. These are the numbers used for the value of Z_m in applying the equivalent circuit analysis or in interpreting data of the type shown in Figs. 3 and 5. Although acoustic losses in solids at high frequencies is not discussed in detail in this chapter this is a property of great practical interest, and an indication of loss in dB/cm for dilatational and shear waves at a specified frequency is included in the next two columns. Values of the attenuation coefficient for the various electrode and bonding layers are not listed. Since the mechanical Q's of these layers are in the range of 10 to 1000, the attenuation in dB/cm at frequencies above 100 MHz could be in the range of 10^2 dB/cm. Fortunately, practical devices use these materials in thicknesses that amount to small fractions of an acoustic wavelength, so that in most practical cases it turns out that the acoustic attenuation due to these layers is negligible. However, the influence on the shape of the loss and phase characteristics of a device due to the impedance mismatch these layers present can be appreciable.[13,14,19] As a result, one of the most important considerations in designing a high-frequency device is to keep the bonding and electrode layers as thin as possible. The last column shows references listed at the end of the chapter.

2. MATERIALS AND TECHNIQUES FOR BONDED PLATE TRANSDUCERS

2.1. Single-Crystal Materials

2.1.1. Piezoelectric Single-Crystal Materials

2.1.1.a. Crystalline Quartz (SiO_2). Crystalline quartz has served as a transducer material for frequencies above 100 MHz for many years. Until the advent of recently developed lapping techniques,[40,41,19] many of the scientific and technological applications of quartz as a high-frequency transducer involved operation of low-frequency plates in overtone modes. When large bandwidth is not an important requirement and high-Q tuning structures can be used with the transducers for purposes of electrical matching, then bonded low-frequency quartz transducers provide a relatively simple inexpensive way of obtaining stress waves in solids at frequencies well over one gigahertz. Examples of this type of approach are contained in Ref. 42 to 47.

Table II lists the constants for the two most important cuts of crystalline quartz used as thickness-mode transducers, the X- and Y-cuts. The AC-cut is also listed; its somewhat lower shear-wave mechanical impedance provides an almost perfect match to fused quartz. The coupling factors for quartz are comparatively low. However, this disadvantage is offset to some extent by the fact that the clamped dielectric constant is low, an especially important advantage for applications requiring large area transducers. The longitudinal and shear mechanical impedances of crystal quartz come close to matching the corresponding impedances of common glass materials used in ultrasonic devices, such as fused quartz and T-40 glass (see Table III).

Along with the advantages related to transducer parameters, crystalline quartz has a number of practical advantages. Among these are low cost, widespread commercial availability (both in the sense of numerous suppliers of the material and in the sense of numerous specialty companies equipped to process it), and excellent mechanical properties that enable the material to be sawed, ground, and polished. In addition, quartz is not water soluble, a point of importance since available bonding methods require standards of cleanliness difficult to obtain without recourse to water-based cleaning methods.

Crystalline quartz occurs in nature in two enantiomorphous forms, referred to as right-handed and left-handed quartz. These two forms are equally prevalent in nature. At 573°C the material has a phase transition going from α-quartz (the low-temperature form) belonging to point group (32) to β-quartz belonging to point group (622). As β-quartz, the material is still useful as a shear-mode transducer for high-temperature operation, but *not* as an extensional-mode transducer.[48] For example, Krause[46] has used Y-cut quartz transducers in combination with gold–indium bonds to operate up to 637°C, which was the temperature at which the gold–indium alloy bond melted.

The relation of the three most useful cuts of quartz for transducer applications to the conventional Cartesian axes associated with the crystal is shown in Fig. 6. An X-cut plate vibrates in a pure thickness-dilatational mode, and the Y-cut and AC-cut plates vibrate in pure thickness-shear

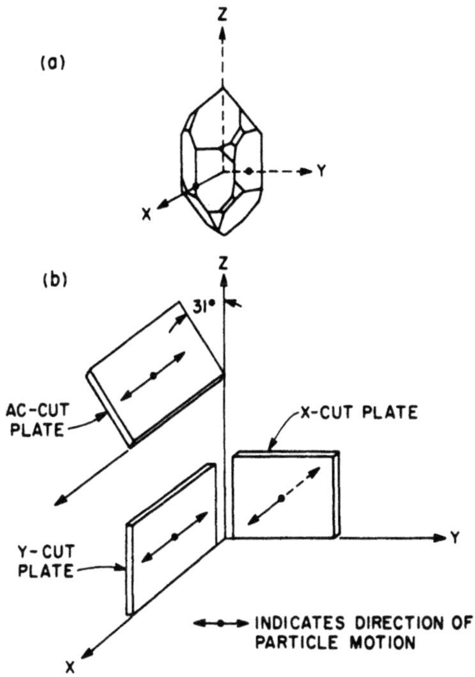

Fig. 6. (a) Right-handed crystalline quartz. (b) Cuts of quartz useful as thickness-mode transducers.

modes. The direction of the particle displacement vectors in the shear wave motions produced by both the Y-cut and AC-cut plates is parallel to the X-axis. The mode purity of these plates can be used to advantage in experiments where the orientational sensitivity of a Y-cut plate or AC-cut plate is used to sense the direction of polarity of a transverse wave motion. Quartz shear-mode transducers have, for example, been applied to studies of Faraday rotation of stress waves in yttrium–iron garnet[49] and to the acoustoelastic effect for shear waves in solids.[47] In the case of the experiments concerned with the acoustoelastic effect, which involve essentially a rotation of the plane of polarization of a transverse wave motion by 90°, suppression of the cross-polarized component by 30 dB was readily attainable.

One important disadvantage of crystalline quartz that at times becomes significant is that there is a relatively large mismatch in the temperature coefficients of linear expansion between crystalline quartz and fused quartz. These materials have been used together in practice for many years, particularly in ultrasonic delay lines. Consequences of this mismatch are the following: (1) Devices using crystalline quartz and fused quartz should, at most, be subjected to processing temperatures within 50°C of room temperature; (2) indium or other materials with some ability to deform plastically or yield are required as the bonding layer for devices that operate over wide temperature ranges; and (3) extremely thin alloy bonds of gold–indium used for experiments at elevated temperatures while useful, as shown by Krause,(46) up to temperatures in excess of 600°C, lose their ability to yield and result in the destruction of the transducer when the sample is returned to room temperature.

2.1.1.b. Lithium Gallate ($LiGaO_2$). Remeika and Ballman(50) have grown lithium gallate in colorless boules of several cubic centimeters volume by the Czochralski pulling technique and reported their initial observations of piezoelectricity in this material. A. W. Warner(51) reported more detailed data for the electroelastic and piezoelectric constants of the material determined from measurements made on resonating plates. The material has several potentially attractive properties including the following: (a) a moderately high coupling factor and mechanical impedance in the longitudinal mode, (b) a high mechanical Q, (c) good chemical stability, (d) moderate hardness, and (e) low dielectric constants.

The material structure is orthorhombic (optically biaxial), of point group (*mm*2). The biaxial properties of the crystal make the axes readily identifiable by optical means. The c crystal axis is the polar axis, and by crystallographic convention, $a < b$. A Cartesian coordinate system associated with the crystal unit cell has X along a, Y along b, and Z along c. The Z-axis is the twofold axis of symmetry and is normal to the plane of the two optic axes. $LiGaO_2$, although piezoelectric with moderately high values of coupling factor, does not appear to be a ferroelectric material. Attempts to observe a Curie temperature below the melting point have been unsuccessful, and it appears that the material is useful as a transducer material to well above 700°C.

Materials in point group (*mm*2) have three plate orientations that result in pure modes of vibration; the Z-cut plate has a pure thickness-dilatational mode and the X- and Y-cut plates have pure thickness-shear modes with very nearly the same constants. Of these three cuts, the potentially most attractive for use in high-frequency transducer applications is the Z-cut. Plates of Z-cut lithium gallate bonded to fused quartz have been lapped to a thickness of 10 μm, corresponding to $f_0 = 300$ MHz, and have shown good transducer characteristics. Figure 7 shows measured insertion loss data(18) obtained from a 300 MHz $LiGaO_2$ transducer along

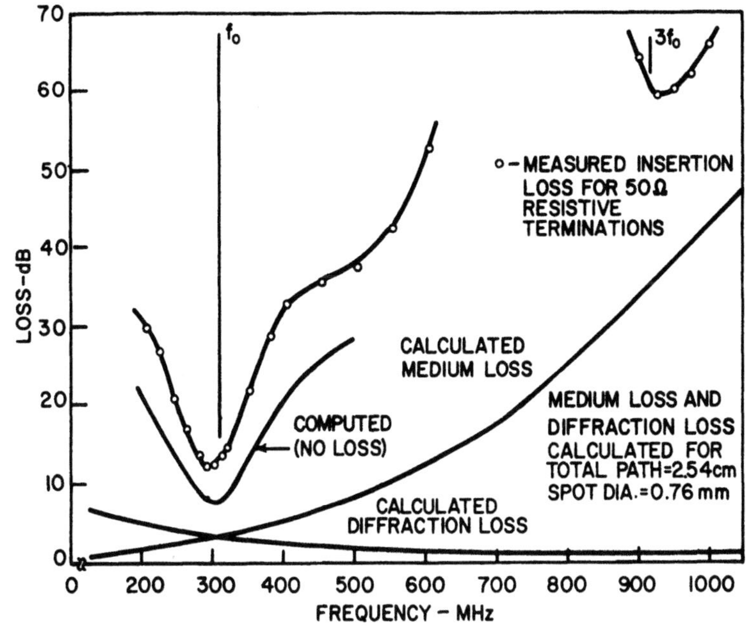

Fig. 7. Loss *vs* frequency response of a bonded lithium gallate, thickness-dilatational transducer 10 μm thick. The graph also shows calculated curves for insertion loss (without medium or diffraction loss), medium loss, and diffraction loss (after Sittig *et al.*[19]).

with calculated curves for medium loss, diffraction loss, and insertion loss with fixed resistive terminations ($R_g = R_l = 1/\omega_0 C_0$).

$LiGaO_2$ is in the same point group as the ferroelectric material barium sodium niobate ($Ba_2NaNb_5O_{15}$); hence they both have the same symmetry properties. $Ba_2NaNb_5O_{15}$ has the advantage of a significantly higher value of k_t, the thickness-dilatational coupling factor. The possible advantages of $LiGaO_2$ over $Ba_2NaNb_5O_{15}$ are (a) $LiGaO_2$ has lower dielectric constants, (b) it is usable over a wider temperature range, and (c) it is easier to process after being grown in the sense that no poling operation is required as is the case with $Ba_2NaNb_5O_{15}$. The differences in the properties of these two materials are primarily attributable to the fact that $Ba_2NaNb_5O_{15}$ is ferroelectric at room temperature and $LiGaO_2$ is not.

2.1.1.c. Bismuth Germanium Oxide ($Bi_{12}GeO_{20}$). Bismuth germanium oxide was first synthesized by A. A. Ballman[52] and many of its crystalline properties have been determined. Below 930°C the material crystallizes in the cubic system in the point group (23); no other phase transitions have been found below this temperature.

The piezoelectric properties of materials in point group (23) are such that none of the simple axial cuts are useful for high-frequency transducers. The useful orientations of $Bi_{12}GeO_{20}$ have been reported,[53] and plate orientations yielding pure modes of vibration are illustrated in Fig. 8. A (111) plate is useful as a thickness-dilatational transducer and a (110) plate is useful as a thickness-shear transducer. The properties of the shear-mode transducer are of greatest interest since the shear coupling factor is 0.235. Only lithium iodate of the presently available single-crystal non-ferroelectric materials has a higher shear coupling factor and lithium iodate has the disadvantage of being water soluble. The clamped relative dielectric constant of $Bi_{12}GeO_{20}$ ($K = 38$) is comparable to that of ferroelectric lithium niobate; however, of these two materials, $Bi_{12}GeO_{20}$ has a significantly lower shear-wave mechanical impedance.

In addition to its transducer properties, $Bi_{12}GeO_{20}$ has several features that make it attractive for use as a high-frequency acoustic medium. Its low melting point makes the growth of fairly large mechanically sound crystals a simpler task than for most high-temperature materials. The

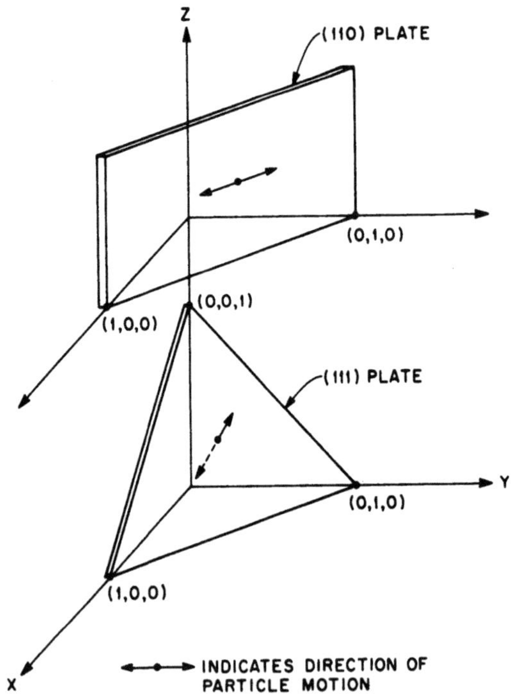

Fig. 8. Cuts of bismuth germanium oxide useful as thickness-mode transducers.

velocities of propagation are relatively low, allowing short crystals to be used to obtain a given delay. There are many directions in which pure modes of propagation are obtained and the properties in these modes are not critically sensitive to orientation. Finally, the shear-wave attenuation in the material is quite low; a mechanical quality factor, Q, of 1.27×10^5 at 500 MHz at room temperature has been reported.[54]

2.1.1.d. Lithium Iodate ($LiIO_3$). Of the single-crystal piezoelectric materials listed in Table II, $LiIO_3$ is the most recent arrival on the scene and the most exciting from the point of view that it combines remarkably high values of coupling factor ($k_t = 0.51$, $k_{15} = 0.60$) with low values of relative dielectric constant ($\varepsilon_{33}^S/\varepsilon_0 = 6$, $\varepsilon_{11}^S/\varepsilon_0 = 8$). The outstanding piezoelectric properties of this material were first pointed out by S. Haussühl[55] who reported 0.95 and 0.80 as the values of the "longitudinal and transverse coupling factors," respectively, although it is not clear that the values reported are for the thickness-mode coupling factors. The values of the transducer constants used in Table II are based primarily on the results of measurements made on small plate-shaped resonators and reported by A. W. Warner *et al.*[56]

Crystals of $LiIO_3$ are hexagonal and belong to point group (6). They are grown in water solution at about 45°C and are colorless. The largest crystals used in the paper by Warner *et al.* were 10 mm long and 3 mm in cross section; the original note by Haussühl reports crystals of 5 cm length and 2.5 cm cross section. Although the crystals are water soluble, no apparent deterioration occurs with exposure to ordinary room atmosphere conditions (relative humidity ~50%); however special cutting, polishing, and cleaning procedures are required so as not bring the material in contact with aqueous solutions.

The melting point of $LiIO_3$ is 420°C and there is a phase transition at 256°C. At room temperature the material cannot be switched and there is no dielectric anomaly below the melting point. Hence the material is not ferroelectric. However, experimental results of the early investigators[55,56] indicate that the electrical conductivity and dielectric constant are strongly varying functions of frequency and temperature. These properties are not yet understood and may eventually be altered by changes in stoichiometry or by doping the material with additives during growth. The difficulties in the dielectric behavior are most important at low frequencies; above 10 MHz the material behaves well as a transducer material with a high coupling factor and low dielectric constant.

At the time of this writing, the electroelastic constants of the material have not been fully determined. Since point group (6) is very similar to point group (6*mm*) the point group of the polarized ceramics, the simple Z-cut and X-cut will probably be the most useful cuts, for thickness-dilatational transducers and thickness-shear transducers, respectively.

2.1.2. Ferroelectric Single-Crystal Materials

In recent years, deliberate attempts have been made to develop improved materials for use in optical devices such as electrooptic modulators and optical frequency doublers.(57) The desirable physical properties for applications of this type are (1) high electric polarizability and (2) low-loss optical transmission. Among the more successful materials resulting from this development effort are synthetic single-crystal ferroelectric materials. Lithium niobate, lithium tantalate and barium sodium niobate are the three materials of greatest interest in the present context. These materials were all originally developed for their potential usefulness in optical devices and later found to be of interest for use as high-frequency piezoelectric transducers.

2.1.2.a. Lithium Niobate ($LiNbO_3$). Large crystals of $LiNbO_3$, 2 cm in diameter and 10 cm long, have been grown by the Czochralski technique of pulling from a melt.(58,59a) This material has the symmetry properties of point group (3*m*) in the rhombohedral system.(59c,d,e) The material melts(58) at 1253°C and has been found to be useful(60) in piezoelectric transducers operating at temperatures up to 1050°C.

Since $LiNbO_3$ is a ferroelectric crystal, domains tend to form as a crystal is pulled from the melt. The domains often appear as concentric annular cylinders in thin sections cut normal to the length of *c*-axis grown crystals.(59a) In growth perpendicular to the *c*-axis, domains tend to occur as flat sandwich structures parallel to the *c*-axis. The presence of domains always degrades the performance of single-crystal ferroelectric materials, both for optical and ultrasonic applications. Fortunately, it is often possible to effect domain reorientation and produce a single-domain crystal by the application of an electric field under proper conditions, a process identified as poling when applied to ferroelectric ceramic materials. The most effective technique for converting multidomain $LiNbO_3$ to single-domain crystals has been found to be by the use of low poling fields at high temperatures. Fields of 0.2 to 5 V/cm have been used for this purpose at temperatures from 1090 to 1210°C. Even at 1200°C, poling has been found to take about ten minutes and longer periods are necessary at lower temperatures.(59b)

When a single-domain crystal is reheated above 1210°C without an electric field applied to the crystal it beomes twinned. Below 1210°C, extended heating with a field present does not affect the domain structure, indicating 1210°C to be the Curie point with an estimated uncertainty of ±10°C. There is no evidence of any other phase change between room temperature and the Curie point. This fact is of considerable practical importance since it accounts for the wide range over which the material is useful as a piezoelectric transducer and, undoubtedly, contributes to the stability of the material when subjected to the temperature excursions and mechanical stresses encountered in transducer fabrication.

Materials in point group ($3m$) have six independent elastic constants, four independent piezoelectric constants, and two independent dielectric constants. Room temperature values of all the electroelastic constants have been determined by Warner, Onoe, and Coquin.[61] Their paper, in addition to reporting experimental and analytical techniques for evaluating these constants, explores the variation of material coupling factor as a function of plate orientation for plates rotated around the principal axes. The four most useful plate orientations are shown in Fig. 9 and the parameters cor-

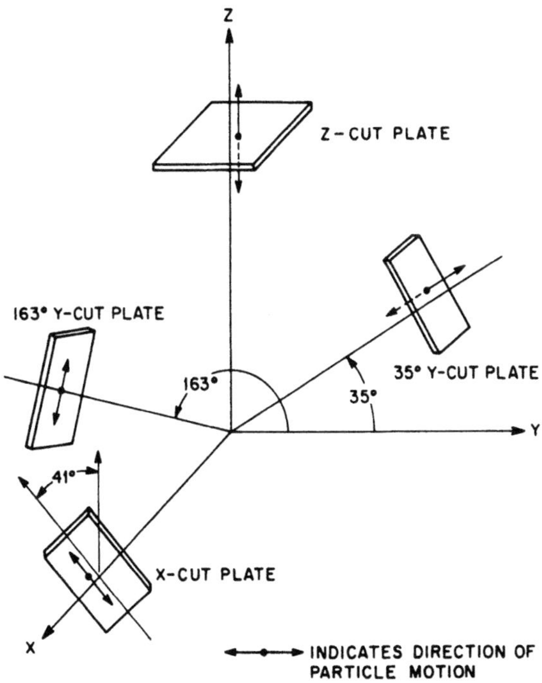

Fig. 9. Cuts of lithium niobate useful as thickness-mode transducers.

responding to these cuts are tabulated in Table II. Two of the cuts are useful as thickness-dilatational transducers and two are useful as thickness-shear transducers.

Because crystals in group ($3m$) possess a low order of symmetry, a detailed study of the electroelastic constants is, in general, needed in order to determine which plate orientations are useful. The criteria used to select the useful plate orientations are the following:

(a) When driven by an electric field parallel to its thickness, the plate must vibrate in a relatively pure mode, i.e., the material coupling factor must be high for the dilatational mode and near zero for the shear, or vice versa.
(b) For a thickness-dilatational transducer, the particle displacement vector must be parallel, within a few degrees, to the plate normal.
(c) For a thickness-shear transducer, the particle displacement vector must coincide, within a few degrees, with the plane of the plate.

The cuts listed satisfy the above requirements. Of the four cuts listed, the Z-cut is a pure mode cut, useful as a thickness-dilatational transducer with a relatively low material coupling factor, 0.17, but with particle displacement exactly normal to the plate. The rotated Y-cuts have particle displacement vectors within approximately 2° of the ideal orientations. In practice, the unwanted mode lies more than 40 dB below the desired mode of transmission when these plates are used as bonded transducers.[(61)] Figure 10 shows data for a 35° Y-cut plate, 10 μm thick bonded to a fused quartz medium.[17]

The X-cut plate of $LiNbO_3$ is particularly interesting from the point of view of application as a thickness-shear transducer since it has an effective coupling factor of 0.68. Two shear modes of thickness vibration are driven by a field applied in the thickness direction; as a result, the particle displacement vector at the surface of the transducer has an elliptical trajectory parallel to the plane of the plate. Fortunately, one shear mode, which will be called the strong shear mode, is excited more efficiently than the other (weak) shear mode. The effective coupling constant is 0.68 for the strong shear and only 0.10 for the weak shear mode.[(61)] Thus X-cut $LiNbO_3$ transducers bonded to a low impedance medium like fused quartz would excite two shear-wave modes of transmission, with one mode approximately 28 dB down from the other at the output (at the frequency of maximum transmission). In contrast to crystalline quartz which has the particle displacement vector along a crystal axis, the particle displacements of the two shear modes are not along the Y- and Z-crystal axes, but are inclined

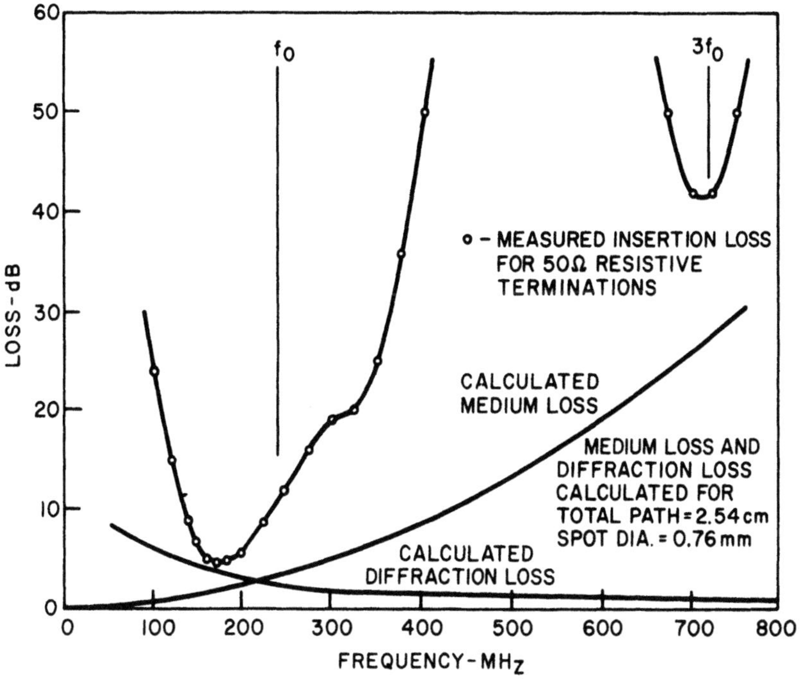

Fig. 10. Loss *vs* frequency response of a bonded lithium niobate, thickness-shear transducer 10 μ thick. The graph also shows calculated curves for medium loss and diffraction loss (after Warner and Meitzler[18]).

to the crystal axes. The direction of the displacement vector for the strong shear mode is 41° from the Z-axis. In high-frequency delay lines involving a body wave mode of transmission in an isotropic medium, this is not an important consideration as long as +X and +Z directions of input and output transducer coincide. However, this fact must be taken into consideration in designing transducers for polygon or strip delay lines which require shear waves polarized in particular directions relative to the major faces of the delay media.

2.1.2.b. Lithium Tantalate ($LiTaO_3$). Lithium tantalate is a member of the same point group as lithium niobate and, in addition to symmetry dependent properties, has many other physical characteristics similar to those of lithium niobate. The growth of single crystals of $LiTaO_3$ by the Czochralski growth technique was reported along with $LiNbO_3$ in a paper by Ballman,[58] and both materials are ferroelectric, as originally reported by Matthias and Remeika.[62] The Curie point of $LiTaO_3$ is 660° ± 10°C and the melting point is approximately 1650°C.[63]

The domain size[63] in the raw (as-grown) crystals of $LiTaO_3$ is much smaller than in $LiNbO_3$. The maximum domain diameter in the as-grown material is of the order of 5 μm compared to several hundred microns in $LiNbO_3$. Further, the domains are antiparallel polar domains.

The temperature coefficient of the resonance frequency for plates of this material is also different. Most cuts of single-crystal materials like $LiNbO_3$ have a negative temperature coefficient of resonance. X-cut plates of $LiTaO_3$ vibrating in the fundamental thickness-shear mode have been found [64] to have a frequency *vs* temperature characteristic that is approximately parabolic in shape with the extremum of the parabola close to 0°C. The temperature coefficient is negative at temperatures below the extremum and positive at temperatures above. The fact that a cut of $LiTaO_3$ has a low temperature coefficient is of considerable significance for resonator applications, but not of particular significance for transducer applications. However, combined with the fact that $LiTaO_3$ is a more efficient electro-optic modulator material, this may result in considerably greater production of $LiTaO_3$ than $LiNbO_3$, which in turn may affect significantly the future cost and availability of the two materials.

The useful cuts of $LiTaO_3$ for transducer applications, along with the room temperature values of all the electroelastic constants, were reported in the paper by Warner, Onoe, and Coquin.[61] Again, the useful cuts for transducer applications and their constants are tabulated in Table II. A comparison of the transducer properties of $LiTaO_3$ and $LiNbO_3$ shows that $LiNbO_3$ has the advantage of combining higher coupling factors (except for the Z-cut plates) and lower mechanical impedance values. For most wideband device applications with media having mechanical impedances of glass-like materials, $LiNbO_3$ will usually be preferred, but $LiTaO_3$ may be useable, even if not preferable, and then cost and availability will become the determining considerations.

2.1.2.c. Barium Sodium Niobate ($Ba_2NaNb_5O_{15}$). Barium sodium niobate, another synthetic ferroelectric material, was grown initially at Bell Telephone Laboratories by L. G. Van Uitert.[65–67] At room temperature, stoichiometric $Ba_2NaNb_5O_{15}$ crystals are orthorhombic, belonging to the pyroelectric crystallographic point group (*mm*2). The melting point of this crystalline material is approximately 1500°C and the Curie point is 560°C. At approximately 300°C, the crystal undergoes a structural transformation. Below 300°C it is orthorhombic; above it is tetragonal, probably belonging to point group (4*mm*).

As in the case of other synthetically grown ferroelectric single-crystal materials, $Ba_2NaNb_5O_{15}$ can be converted to single domain by application of a poling field at a temperature a few degrees below the Curie point.[67] This material is different, however, from $LiNbO_3$ and $LiTaO_3$ in that a phase transition takes place at a temperature between room temperature and the Curie temperature. This structural transformation ordinarily produces a high degree of microtwinning in the as-grown crystal. The twinning is such that all crystallites have their *c*-axes in the same direction but the *a*- and *b*-axes are interchanged. The microtwinning may be removed by heating the crystal above 300°C with a compressive stress applied along one of the tetragonal *a*-axes.[67] The crystal is cooled under stress through the transition temperature of 300°C, and the axis along which the stress is applied becomes the intermediate or *a*-axis. Once detwinned in this manner, a crystal of $Ba_2NaNb_5O_{15}$ is a uniform single crystal and can be handled, cut, and polished near room temperature by normal processing techniques without undue concern that the crystal will revert to the twinned state.

Single crystals of $Ba_2NaNb_5O_{15}$ have shown outstanding performance in nonlinear optical devices such as frequency doublers and parametric oscillators, as well as in electrooptic modulators. Compared to $LiNbO_3$ and $LiTaO_3$, they have shown a superior ability to withstand intense laser radiation.[67] The remarkable usefulness of this material in optical device applications holds forth the promise that it will become commercially available in large-size crystals at moderate cost.

The potential usefulness of $Ba_2NaNb_5O_{15}$ as a high-frequency thickness-dilatational transducer material was reported by Warner *et al.*,[68] and a detailed determination[69] of the electroelastic constants has been carried out. The constants of this material relevant to transducer operation are given in Table II. For dilatational-mode transducers, this material offers the advantage of a high coupling factor, 0.57, combined with a relatively low value of K (= 32). Other advantages are a lower dilatational mechanical impedance than those of $LiNbO_3$ and $LiTaO_3$ and the fact that the high coupling factor cut is a simple Z-cut rather than a rotated Y-cut. In addition, measurements of coupling factor over the temperature range between 0° and 100°C have shown that the coupling factor is constant in this range. As of 1970; the value of $k_t = 0.57$ was the highest value known for a single-crystal material usable at and above room temperature.

Because of the fact that the microstructure of the twinned regions in crystals that are not single domain involves an interchange of the *a*- and *b*-axis directions, a crystal that is poled but not detwinned has a value of k_t essentially the same as that of a single-domain crystal. In a sense, a twinned but poled crystal acts like a poled ferroelectric ceramic which has Z-axis alignment but random orientation of the other axes. Thus the detwinning operation, which is absolutely essential when the material is used in optical device applications, is not essential for application of the material in piezoelectric transducers.

Table II also shows the constants for thickness-shear operation. Since the *a*- and *b*-axes of the unit cell differ in length by only three parts in 1700, it is not surprising to see that the constants for X- and Y-cut plates are almost the same. Because the relative dielectric constants for X- and Y-cuts are large, 222 and 227 respectively, and the coupling factor small (less than half of k_t), the material does not appear to be attractive for use in high-frequency thickness-shear transducers. All the other piezoelectric materials listed in Table II have shear coupling factors, for some orientation, greater than the thickness-dilatational factor; but this inequality does not apply for $Ba_2NaNb_5O_{15}$.

2.2. Ceramic Transducer Materials

2.2.1. Comparison of SPN and PZT-7A Ceramics

At the present time there are two piezoelectric ceramic materials commercially available and suited for use as high-frequency transducer media. These materials are a sodium–potassium niobate ceramic[70,71] (SPN) and a lead zirconate–lead titanate ceramic[72] (PZT-7A).* Both materials combine high values of coupling factors with values of dielectric constants that are low compared to most other ceramic materials. An additional important characteristic, and one that is an advantage for high-frequency applications, is that both materials possess fairly high values of frequency constant. The properties of these materials of interest for transducer applications are given in Table II.

As can be seen from Table II, the two ceramics have nearly equal shear and dilatational mechanical impedances. This enables the figure of merit M, defined earlier by Eq. (9) to be used in comparing their performance capabilities. In order to use directly the quantities tabulated in Table II, a revised figure of merit, M_1, is defined here that is directly proportional to the original M and that is given by

$$M_1 = kf_0t/K \tag{12}$$

* SPN is marketed by Bausch and Lomb, Inc. and PZT-7A by Vernitron, Inc.

Table IV. Figures of Merit for SPN and PZT-7A Thickness-Mode Transducers

Material	M_1 (dilatational) MHz · μm	M_1 (shear) MHz · μm
SPN	5.15	2.24
PZT-7A	5.10	1.82

The values of M_1 for the thickness-dilatational and thickness-shear modes are shown in Table IV.

Use of the "figure of merit" concept in this case makes obvious one of two basic conclusions about the relative usefulness of these two materials. This conclusion is that for dilatational-mode applications the two materials are essentially equivalent. A second conclusion (more evident from Table II than from Table IV) is that, for shear-mode applications, the SPN has a definite advantage over the PZT-7A material resulting primarily from the fact that the SPN material has a value of shear wave velocity almost 50% greater than that of PZT-7A. This becomes of greater significance in the use of high-frequency transducers than is implied by the difference in the values of M_1 because for extremely thin transducers, every additional micron of thickness that a plate may have in reaching a frequency above 100 MHz increases disproportionately the probability of realizing a finished transducer.

High-frequency transducer applications place conflicting requirements on ceramic materials. Transducers in the range of 100 to 200 MHz have thicknesses in the range of 20 to 10 μm. On the one hand then, it is desirable to have the grain size of about 1 μm in order to have the transducer 10 or more grains thick. Larger grain size material might be expected to suffer from pull-outs (voids) and excessive crumbling when polished to dimensions of the order of 10 μm. On the other hand, it is difficult with present techniques of producing ceramics to retain strong ferroelectric behavior (and hence large coupling factors) when the grain size becomes smaller than a few microns. Electron photomicrographs of the PZT and SPN ceramics indicate it is possible to achieve a maximum grain size in the range of 1 to 5 μm, and in this range the high coupling factors indicated in Table II are attainable. The problem of retaining strong ferroelectric behavior for grain sizes under 1 μm imposes an intrinsic limitation on ceramic materials, restricting their usefulness to frequencies below a few hundred megahertz.

At this point, it should be noted that although PZT-7A has properties which are attractive for high-frequency use, there does not appear to be any mention in the literature of the use of this material at frequencies above 100 MHz. On the other hand, ultrasonic delay lines using SPN in the shear mode have been reported[1] operating at frequencies above 100 MHz. Since the frequency constant for the thickness-dilatational mode is approximately twice that of the thickness-shear, this frequency of operation implies that frequencies near 200 MHz should be attainable in the thickness-dilatational mode. Although ceramic transducers can be lapped thin enough to work at frequencies of 200 MHz, this approach to realizing high-frequency transducers is not attractive because the electrical impedance of transducers with useful areas would probably be so low as to cause practical difficulties.

2.2.2. Considerations Related to the Ferroelectric Properties of Ceramics—Factors Influencing Poling and Depoling

Ferroelectric ceramics are polycrystalline materials formed by reacting together finely powdered mixtures of metallic oxides (sometimes carbonates) at high temperatures and, in some cases, high pressures (hot-pressing). Even though the crystallites constituting the ceramic body formed by the reaction may be as small as a few microns in diameter, the material within a single crystallite, when cooled below room temperature, still possesses many of the properties of a multidomain ferroelectric crystal. (The Curie point is approximately 350°C for PZT-7A and 420°C for SPN.) Immediately after formation, a ferroelectric ceramic material has a net remanent polarization of zero as a result of the random distribution in the direction of the dipole moments within the domains. However, these materials can be given a permanent dipole moment and a strong piezoelectric response by the application of a dc field, usually for times on the order of minutes, in a temperature range between 25 and 300°C. This process is referred to as "poling."

The piezoelectric properties of poled ceramic materials are, as a rule, not as stable as those of single-crystal nonferroelectric materials. The poling process introduces strains into the ceramic material which are gradually relieved by relaxation mechanisms involving predominantly domain wall motion. However, materials and techniques have been developed which result in devices with sufficient stability to make them useful in practical applications. Nonetheless, a large part of the engineering problem involved in developing devices using these materials resides in the development of processing techniques that result in realizing high values of coupling factors along with stable performance characteristics.

Since SPN is, at present, the only ceramic material for which information regarding use at frequencies above 100 MHz is available, the remainder of this section will concentrate on the techniques used to form high-frequency SPN transducers.

2.2.2.a. Poling Procedures for SPN High-Frequency Transducers. The standard procedure for poling many ceramic materials, including SPN, is first to apply poling electrodes to the sample. The electrodes may be layers of material such as air-drying silver paste or electroless nickel. The sample is then immersed in a bath of silicone oil and heated to approximately 200°C. A voltage is applied to establish a field of 2 V/μm and maintained for a period of about 10 min. With this voltage still applied, the bath is cooled to below 100°C. At this temperature the sample can be removed and washed in a suitable poling oil solvent, such as trichloroethylene.

The poling technique using an oil bath is widely used and generally quite satisfactory for large size samples. However, ceramics are slightly porous and there is some tendency to soak up oil. Oil absorbed into the ceramic can be a source of difficulty in attaining good electrode adherence, particularly if the ceramic is poled in thin sections as is often done for high-frequency transducers. To avoid contamination of the ceramic by poling oil, Sittig and Cook[(1)] have used a technique in which the ceramic is poled in a chamber that heats the sample to the poling temperature and immerses it in a nitrogen atmosphere under a pressure of 20 lb/in^2. The nitrogen gas under pressure has been found to carry out quite well the two basic functions of the poling bath: (1) the transfer of heat to the sample being poled by conduction and (2) the prevention of arcing or electrical breakdown that may occur between electrodes when a ceramic is poled in air at atmospheric pressure.

2.2.2.b. Depoling Caused by Elevated Temperatures. In the fabrication of devices using ceramic transducers, it is necessary to use bonding and lapping techniques that do not subject the poled ceramic sample to excessive increases in temperature. Figure 11 shows the rate at which elevated temperatures cause SPN to depole.[73] The data were taken by placing a transducer with a known coupling factor in an oil bath at the specified temperature for 5 minutes, then removing it and remeasuring the transducer. The transducer was then put back into the oil for three more 5 minute periods, and the coupling factor was measured after each period. The data

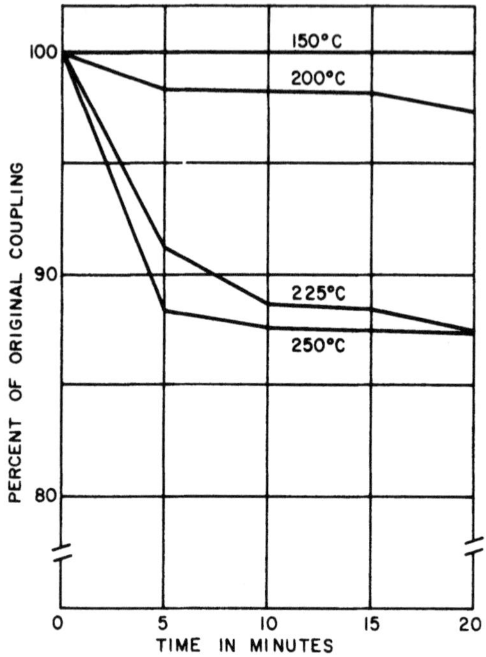

Fig. 11. Loss in coupling factor of SPN transducers as a function of time immersed in an oil bath at elevated temperatures (after Dean[73]).

indicate that significant changes in coupling factor occur in short times if the ceramic is subjected to temperatures above 200°C.

2.3. Bonding and Lapping Techniques for High-Frequency Transducers

The ability to fabricate ultrasonic devices with low insertion loss and wide bandwidths at frequencies above 100 MHz using bonded plate transducers rests in large measure upon the development of improved techniques for bonding transducer materials to substrates and polishing these bonded plates to extremely thin layers.

At present, there are two main bonding techniques; one using transparent organic materials to form a thin adhesive bonding layer and the other using combinations of metal films, with indium as the principal metal, joining two metallized surfaces together.

2.3.1. Epoxy Bonding Techniques

Epoxy resin compounds are the most useful of the organic materials for realizing bonds of good stability over temperature ranges between 0 and 80°C. The specific mechanical impedance of these materials is low relative to other solid materials, as can be seen from Table III. Analysis[13] has shown that in order for the low mechanical impedance of a bonding layer of this type not to have a serious effect on the bandpass characteristics, it is necessary for the thickness of the bonding layer to be of the order of $0.01\lambda_b$, where λ_b is the wavelength of sound in the bonding layer. For acoustic waves in epoxy at a frequency of 100 MHz, this corresponds to 0.26 μm for dilatational waves and 0.12 μm for shear waves. To realize such thin bonds in practice with a high rate of yield requires extreme care, working in dust-free environments with solvents and bonding materials filtered to remove all particulate matter larger than a few tenths of a micron.[74]

An extremely important aid to the realization of thin organic bonds has been the development of a bonding technique[75] that incorporates into the transducer–bond–substrate combination a "built-in" interferometer. The basic configuration for doing this is shown in Fig. 12. The use of a half-transparent film on one side and a fully reflecting film on the other side of the bond forms a means of generating constant-thickness fringes that can be observed, under proper lighting conditions, either directly by eye or with the aid of a microscope. Examples[76] of the appearance of

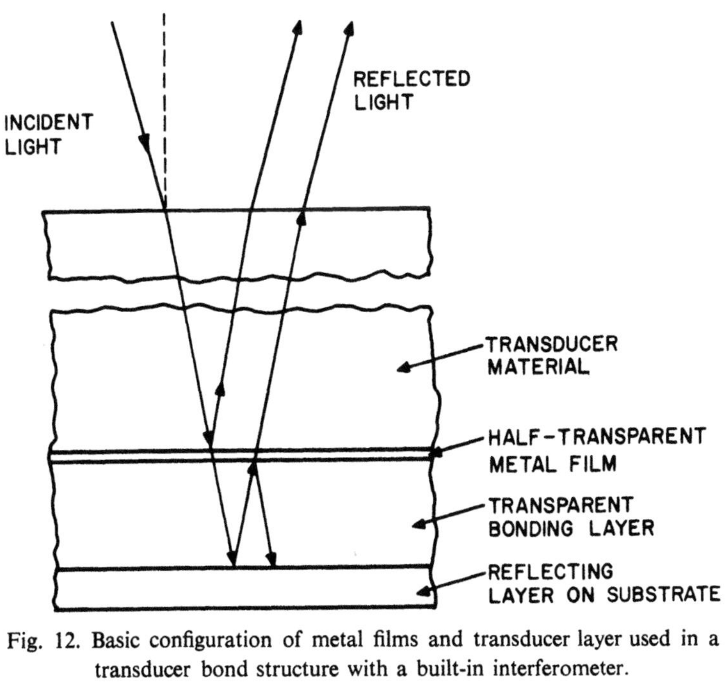

Fig. 12. Basic configuration of metal films and transducer layer used in a transducer bond structure with a built-in interferometer.

"clean" and "dirty" bonds as made observable by this technique are shown in Fig. 13. Details of the techniques used to form epoxy bonds and the interpretation of the interferometric patterns are given in a paper by J. S. Jones.[74]

It should be borne in mind that most of the techniques developed in connection with the epoxy bonds are equally applicable to other organic

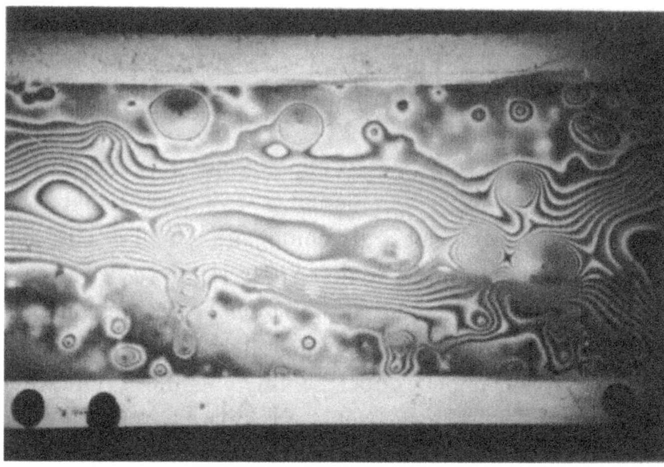

(a) DIRTY BOND

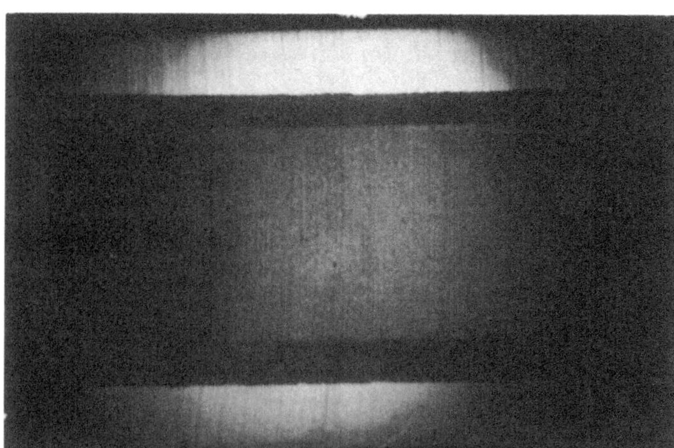

(b) CLEAN BOND

Fig. 13. Appearance of fringe patterns in the case of a dirty bond and an exceptionally clean epoxy bond as seen with sodium light and low-power (~2×) magnification (after Sittig[76]).

materials such as salol (phenyl salicylate) and phenol (phenyl benzoate). These materials are most useful in experimental work because they have relatively low melting points that make possible the recovery of transducers bonded with these materials. Although a solvent for the epoxy bonds does exist, epoxy bonded transducers, particularly if the layer is extremely thin, are almost impossible to recover intact.

2.3.2. Indium Bonding Techniques

Indium has been used as a bonding agent in joining together glass and crystalline materials for some time. Discussion of indium bonding techniques appears in a number of publications[6–8,10,77] related to ultrasonic devices. Recent refinements of the basic techniques related to devices operating at frequencies above 100 MHz have been reported by collaborators at Bell Telephone Laboratories.[1,19] The refinements have been directed along lines of producing the thinnest possible layers with good adherence by metallizing the transducer and substrate surfaces and bonding in a vacuum chamber, and by making the indium bond at room temperature in order to avoid the problems arising from different rates of thermal expansion for different materials in the transducer–bond–substrate combination.

The basic sequence of operations followed in preparing an indium bond is given[19] as follows:

1. The transducer blanks and acoustic medium are prepared by crystallographic orientation (where needed). Transducers are lapped to a thickness convenient for bonding (not necessarily the final thickness). All surfaces to be bonded are polished flat.
2. The metal layers necessary for adhesion, electrical conduction, and alloying with the indium are deposited on the surfaces to be bonded.
3. Thin indium layers are deposited over the previously vapor plated metal layers on the two surfaces to be bonded. Immediately after the surfaces are coated with indium they are forced together under a pressure of 1,000 to 10,000 psi by a pneumatic press inside the vacuum chamber.
4. The thickness of the bonded transducer is reduced to the value required for the desired frequency of operation.
5. Outer electrode layers are vapor-deposited on the transducer.

At the present time, the combination of layers most useful in bonding a wide variety of crystalline and glassy media is a flash of chromium ($\sim$20 nm) followed by a thicker layer of gold ($\sim$100 nm). A relatively thin layer of indium (~50 nm) is applied to the surfaces immediately before joining. Although this bond is made at room temperatures, the indium and gold combine[46] to form a bond usable to 600°C. The electrical and mechanical properties of this bond are quite good; however, the impedance mismatch of the gold layer influences transmission properties by significantly reducing the total bandwidth.[13,19] For this reason other combinations of metals such as silver and indium are under investigation[19] in order to find a satisfactory combination of materials to achieve 3 dB bandwidths of several hundred MHz.

2.3.3. Transducer Thinning Techniques (Lapping and Sputter Machining)

The techniques of bonding described in the two preceding subsections enable polished plates of transducer material to be joined to polished surfaces of an acoustic medium with intervening layers of a few hundred nanometers. The prevention of the inclusion of particulate contaminants is important, first, in forming the thin bonding layers and, second, in reducing the risk of fracture when subsequently lapping the transducer to thicknesses of the order of 10 μm.

The achievement of transducer layers with lateral dimensions of several mm and thicknesses of several μm, such as those shown in Fig. 1, requires elaborate means to measure and maintain alignment of surfaces with high angular precision of the order of ± 5 sec of arc. High precision spectrometer tables and autocollimators are available that can supply the required accuracy. The jigs that are used to carry the medium on which the transducer is bonded must be capable of fine adjustment to enable the outer surface of the transducer to be lapped parallel to the substrate. Several different designs suited to accomplishing this have been described.[19,40,41]

Lapping and polishing of the transducer materials can be satisfactorily carried out on layers of tin charged with micron-size diamonds.[19] The metal lapping surface offers the advantage that it can be readily flattened or reconditioned by cutting it on a lathe. Diamond lapping pastes of good uniformity covering the range from 0.1 to 15 μm in a number of steps are commercially available. The paste forms a water-soluble lubricant slurry which is kept to a minimum in order to minimize damage to the substrates. As the thickness of the transducer is reduced, it is necessary to make frequent checks of the thickness of the layer and the parallelism of the surface being lapped with the reference surface. In the technique[19] developed at Bell Telephone Laboratories, a Zeiss Light-Section microscope is used to check thickness; the basic measuring capability of this instrument is ± 0.5 μm for samples less than 25 μm in thickness. Parallelism is checked with a Wild spectrometer table; this instrument is capable of measuring to ±1 sec of arc.

In addition to the lapping techniques, which basically involve refinements of techniques that have been used in the quartz resonator and optical lens industries for many years, the technique of "sputter machining" layers of transducer materials to dimensions of a few microns has recently been investigated. In sputter machining, material is removed by bombarding the transducer layer with high-energy ions drawn from a gaseous plasma. Experimental results have been reported[78] for the use of this technique on sputter-deposited zinc oxide films as well as on bonded single-crystal Y-cut lithium niobate and Y-cut quartz. The highest frequency attained was a half-wavelength resonance frequency of 1.8 GHz corresponding to a thickness of about 2 μm. A disadvantage associated with this technique at present is the tendency to remove material at a faster rate from the edges of the piezoelectric layer than from the center. This causes the usable area of the transducer layer to decrease as the layer is made thinner. Advantages associated with the technique are that transducer layers in the micron thickness range can be prepared using relatively simple apparatus and relatively unskilled personnel.

3. EVAPORATED AND SPUTTERED FILM TRANSDUCERS

3.1. Electroelastic Properties of Cadmium Sulfide and Zinc Oxide in the Form of Thin Films

At present the two most important materials by far for application as thin film transducers are CdS and ZnO. Both compounds exist in bulk forms as solids with hexagonal structures. Although CdS also exists in a cubic form, only the hexagonal form has been commonly used in films for high-frequency transducers.[16]

The physical properties of CdS and ZnO as bulk solids have been discussed in the earlier chapter by D. Berlincourt. The parameters of particular interest for applications as high-frequency transducers are given in Table II of this chapter, as determined from measurements on bulk samples. There is, in general, good agreement between bulk values and thin film values. However, the coupling factor for films may be significantly lower than for the bulk material since, unless epitaxially grown, the films are collections of grains with only limited alignment of crystalline axes.

Coupling factors in films as produced by present techniques[16,79] are commonly in the range of 60 to 90% of the bulk single-crystal values; although one publication[12] has reported a value of the longitudinal-mode coupling factor for CdS films equal to the bulk value.

The fact that polycrystalline films of CdS and ZnO can have coupling factors almost equal to that of the bulk single-crystal material is related to the symmetry properties of the point group to which they belong. Both CdS and ZnO are members of point group ($6mm$). Materials in this point group have *both* electric and elastic symmetry in the plane perpendicular to the *c* axis. As a result, a polycrystalline layer that has some preferential alignment of the *c*-axis directions of individual grains behaves like a single-crystal layer even though the grains are randomly oriented with respect to their *a*-axes.[80]

Insensitivity to random orientation of a-axes is one important property of CdS and ZnO films that contributes to their usefulness. Another property exhibited by both materials is that there is a distinct preference for the c-axes of grains to orient roughly parallel to the vapor beam with which they are deposited, with the plus (metal ion) layer toward the beam under some conditions. In early work[21] with CdS normal incidence was used and the films deposited had the c-axes approximately normal to the surface. Later experiments by Foster[81] demonstrated that efficient shear-mode transducers could be formed by depositing layers with the beam incident at an oblique angle. The technique of controlling the orientation of the layer by the angle of incidence of the vapor beam has been most effective with CdS transducers. With ZnO transducers, a sputtering technique rather than an evaporation technique is most frequently used and while control of the angle of incidence is important it is not the sole determining factor,[79] as will be discussed in more detail in a later subsection.

The ability to achieve close to single-crystal performance in polycrystalline films is only one consequence of the symmetry properties of ($6mm$) materials. There are several other important properties directly relevant to transducer applications:

(1) A pure dilatational mode of vibration can be driven by a field in the thickness direction when the *c*-axis is normal to the plane of the layer.
(2) A pure shear mode of vibration can be driven by a field in the thickness direction with *two* different layer orientations: (a) when the *c*-axis of the film is in the plane of the layer and (b) when the *c*-axis is tilted about 40° away from the normal to the layer.[16]

The particle displacement vector is parallel to the c-axis when the c-axis is the plane of the layer and parallel to the projection of the c-axis on the layer when the axis is inclined to the normal.

(3) A film with an average c-axis direction normal to the layer can be deposited directly onto a dielectric single-crystal substrate and in some cases used to drive individually (by means of a re-entrant cavity technique) all three independent modes of propagation along a particular axis of the crystal.[82] To excite the dilatational mode, the field is applied in the thickness direction. To excite the shear mode, a cavity must be used that applies an electric field vector in the plane of the film. The shear strain developed in the film will have a particle displacement vector parallel to the electric field vector. By proper orientation of a single-crystal substrate relative to the direction of the applied electric field, either shear mode in an anisotropic crystal along the direction normal to the film can be excited independently. This shear-mode drive technique can be applied to polycrystalline films of ($6mm$) materials as well as to single-crystal films grown by epitaxial growth techniques.[80]

The above list of properties of ($6mm$) materials indicated that there are two layer orientations providing a pure thickness-shear vibration and one layer orientation providing a pure thickness-dilatational vibration. Plate or film orientations other than these will have piezoelectric properties such that the transducer will vibrate in a mode of vibration involving both shear and dilatational component waves. Analysis of this mixed condition has been made[(16)] in terms of effective coupling factors $k^{(S)}$ and $k^{(D)}$ for the shear and dilatational components as a function of plate orientation angle φ. The results are shown in Fig. 14 for CdS. The curves calculated for ZnO behave in a very similar fashion.[(16)]

3.2. Evaporation Techniques for Forming CdS Transducers

There are two basic vacuum evaporation techniques for forming CdS film transducers. One technique[(20,83)] employs an indirect evaporation system in which individual boats of cadmium and sulfur are heated and in which the substrate to be covered is separated from the sources by a baffle. The other basic technique[(21)] employs a direct evaporation system in which a single CdS source is used to supply a beam of cadmium and sulfur atoms directed towards the substrate to be covered. A third technique, which has been reported[(84)] but which appears to be little used, is one in which CdS is vaporized by electron bombardment of a pellet of the material.

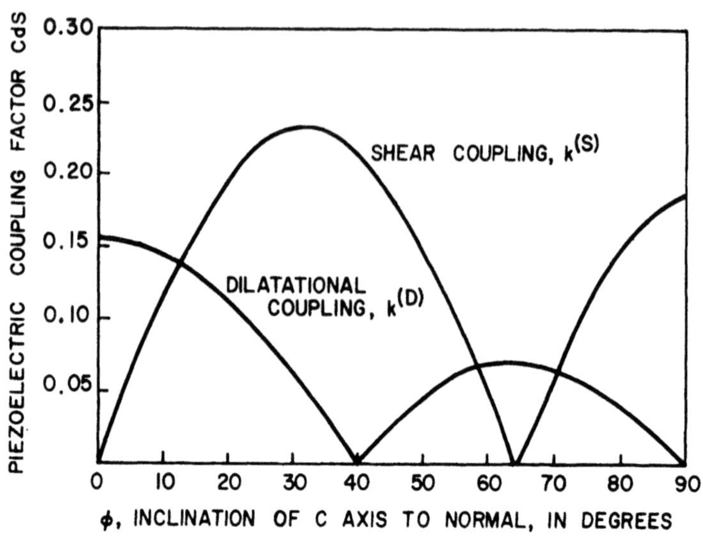

Fig. 14. Variation of effective shear and dilatational coupling factors as a function of the angle φ (after Foster *et al.*[16]).

The problem of stoichiometry is common to all techniques of forming film transducers. In the case of CdS films, the temperature required to vaporize the CdS is in excess of the decomposition temperature. As a result, the vapor in the vacuum system is in all cases composed of the elemental species. Films grown under conditions of equal cadmium and sulfur fluxes, as in the case when they are evaporated from a single CdS source, are usually sulfur deficient because there is a greater tendency for the sulfur to evaporate from the surface of the substrate than there is for the cadmium. Films with excess cadmium are semiconducting and this condition results in inefficient action as a piezoelectric transducer. The film stoichiometry can be improved by heating the substrate to reduce the time for which free cadmium resides on the substrate surface and by reducing the deposition rate. Under these conditions only the compound, CdS, sticks because of its higher sublimation point. Another means of improving film stoichiometry is by including an additional source of sulfur in the system which is heated independently of the source containing the main supply of CdS.

The resistivity of thin films is an important consideration and, as just indicated, the stoichiometry of the film is one important factor influencing resistivity. Another important factor is the material used as the electrode layers. The resistivity of a CdS film can be increased by using suitable monovalent dopants (e.g., Ag or Cu) acting as electron acceptors. In some cases, the silver or copper has been used as an electrode layer and in other cases it has been introduced in the film by coevaporation from a separate source.[21]

3.2.1. CdS Films by Indirect Evaporation

Figure 15 shows a vacuum chamber arranged for the formation of CdS film transducers by the technique of indirect evaporation. Powdered cadmium and sulfur are placed in separate containers. A baffle is located between these sources and the substrate on which the film is to be deposited. The assembly is located inside a heated enclosure that is used to maintain a substantial vapor pressure of cadmium and sulfur. In addition to the deposition on the substrate surface, deposition takes place on all the other surfaces at low enough temperatures, including the quartz crystal deposition rate monitor. Typical conditions are as follows: sulfur temperature 150°C,

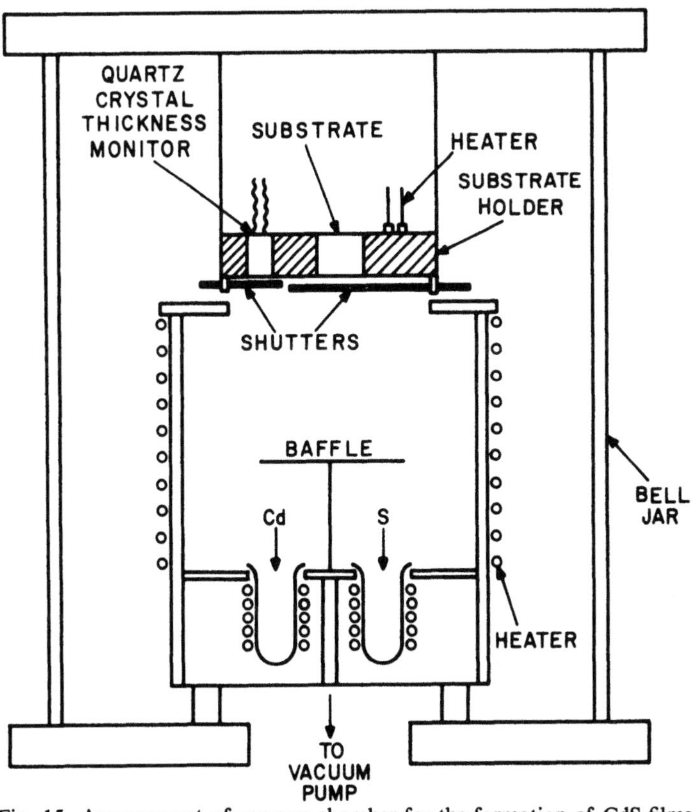

Fig. 15. Arrangement of vacuum chamber for the formation of CdS films by the technique of indirect evaporation (after de Klerk[83]).

cadmium temperature 400°C, inner chamber wall temperature 160°C, and substrate temperature 200°C.

The system described here was originally developed by de Klerk and Kelly[20,83] and used to deposit ZnS films, as well as CdS, by the substitution of zinc for cadmium. One advantage of this system is that film growth takes place under close to equilibrium conditions that favor, in the presence of a suitable single-crystal substrate, epitaxial growth; that is, formation of a single-crystal film rather than the type of polycrystalline film usually produced. The technique of indirect evaporation yields *c*-axis films and is not suited to making films with the *c*-axis inclined at about 40° to the normal or parallel to the plane of the plate as is desirable for shear-mode films used in high-frequency ultrasonic delay lines.

3.2.2. CdS Films by Direct Evaporation

A complete evaporation station for producing film transducers is shown in Fig. 16. The substrate is placed in a tiltable holder that supports the sample over the source and enables the sample to be inclined so that the vapor beam can be directed onto the substrate face at any desired angle. This arrangement allows one to obtain films with the *c*-axis other than normal. The source of CdS is usually used along with a separate source of sulfur and both sources are housed in a water-cooled enclosure equipped with a solenoid-operated shutter.

The ability of CdS to form a layer with the *c*-axes of crystallites oriented along the direction of the vapor beam is in part dependent upon the nature of the metal electrode layer. Foster has found that the best deposition of inclined axis films is obtained with randomly oriented substrate films of gold. In this case, even though the first 100 nm or so of layer growth is oriented with the *c*-axis approximately normal to the substrate, as the thickness of the layer increases, the *c*-axis becomes progressively inclined towards the incident vapor beam reaching full alignment at a thickness of about 1 μm. Studies of the influence of the substrate temperature, the rate of deposition, and the angle of evaporation have been reported in the literature,[16,83,85] and a detailed list of references treating this specialized topic can be found in a recent review chapter by N. F. Foster.[22]

3.3. Sputtering Techniques for Forming ZnO Transducers

The evaporation of zinc oxide is much more difficult than cadmium sulfide because there is a much greater difference in the vapor pressures of

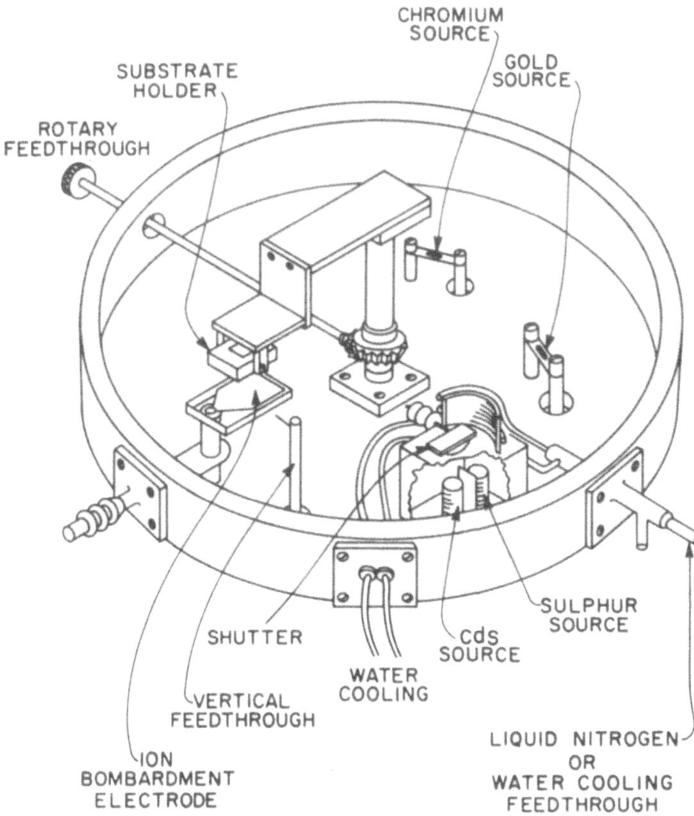

Fig. 16. Arrangement of vacuum chamber for the formation of CdS films by the technique of direct evaporation (after Foster).

the two components. Even so, successful formation of ZnO film transducers has been performed by use of the technique of reactive evaporation.[86] In this technique, zinc is evaporated into an oxygen-containing atmosphere; but to obtain these films it is necessary to cool the gold plated substrate to a temperature in the range 0 to -50°C. Oxygen is introduced into the vacuum system through a controlled leak to maintain a pressure of 10^{-3} to 10^{-4} torr during the evaporation.

The difficulties associated with a direct evaporation technique of ZnO account for the fact that this technique was not the first means used to achieve piezoelectrically-active films of ZnO. Piezoelectrically-active ZnO films were first achieved by Wanuga *et al.*[87] with a reactive diode sputtering technique using zinc as the target material and an oxygen-argon mixture at 50 μm pressure. Shortly thereafter, Rozgonyi and Polito[88] reported successful diode sputtering of ZnO films with ZnO as the target material. In this work, deposition rates of up to 50 nm/min were reported on substrates heated to about 200°C.

The deposition of ZnO films by means of sputtering techniques is by far the more successful approach in practice. In the process of sputtering, high-velocity particles impinge on a target material and remove atoms. These atoms deposit as thin films on surrounding surfaces. Practical sputtering systems normally use ions as the impinging particles. The removal of material from the target occurs during an electrical discharge in a gaseous plasma which provides the ions for acceleration to the target to produce the desired action. In diode sputtering, the necessary plasma can be generated by a dc glow discharge through application of a high potential between two electrodes at approximately 2 to 10×10^{-2} torr gas pressure. Diode sputtering works best with source materials that are at least slightly electrically conductive since otherwise the process is interrupted by the field produced by the ions which accumulate on the target. This can be avoided by the use of high-frequency fields which permit sputtering all types of electronic materials (conductors, semiconductors, and insulators).

Figure 17 shows several features of a diode sputtering facility developed by H. D. Cook[89] for the fabrication of ZnO films. The sample on which the film is deposited is suspended from a horizontal plate attached to a centrally-located rotatable turret. This arrangement is used to enable the sample to be positioned above the sources used to vapor plate metallic substrate films before the sample is positioned over the ZnO cathode. The cathode is a hot-pressed ceramic disk of ZnO supported by a water-cooled structure. The cathode is given a negative potential of approximately 2000 V which sustains a discharge in an argon–oxygen atmosphere at a pressure of approximately 10^{-1} torr. Under these conditions a deposition rate of 0.5 μm/h is obtained. During the film deposition the temperature of the sample substrate is monitored and held at 25°C by controlling the high voltage. Heat developed in the sample and its surroundings is removed by means of a heat sink placed in contact with the horizontal rotatable support plate during the film deposition process. The arrangement depicted in Fig. 17 has been used to deposit films as thick as 20 μm. Experimental data obtained for a 3.2-μm-thick ZnO film is shown in Fig. 18. This film was deposited on fused quartz and had inner and outer electrode layers of Cr–Au, with approximately 50 nm of Cr and 100 nm of Au used to form each electrode. The outer electrode was a circular spot 0.50 mm in diameter. Figure 18 shows the insertion loss measured for 50 Ω resistive termination

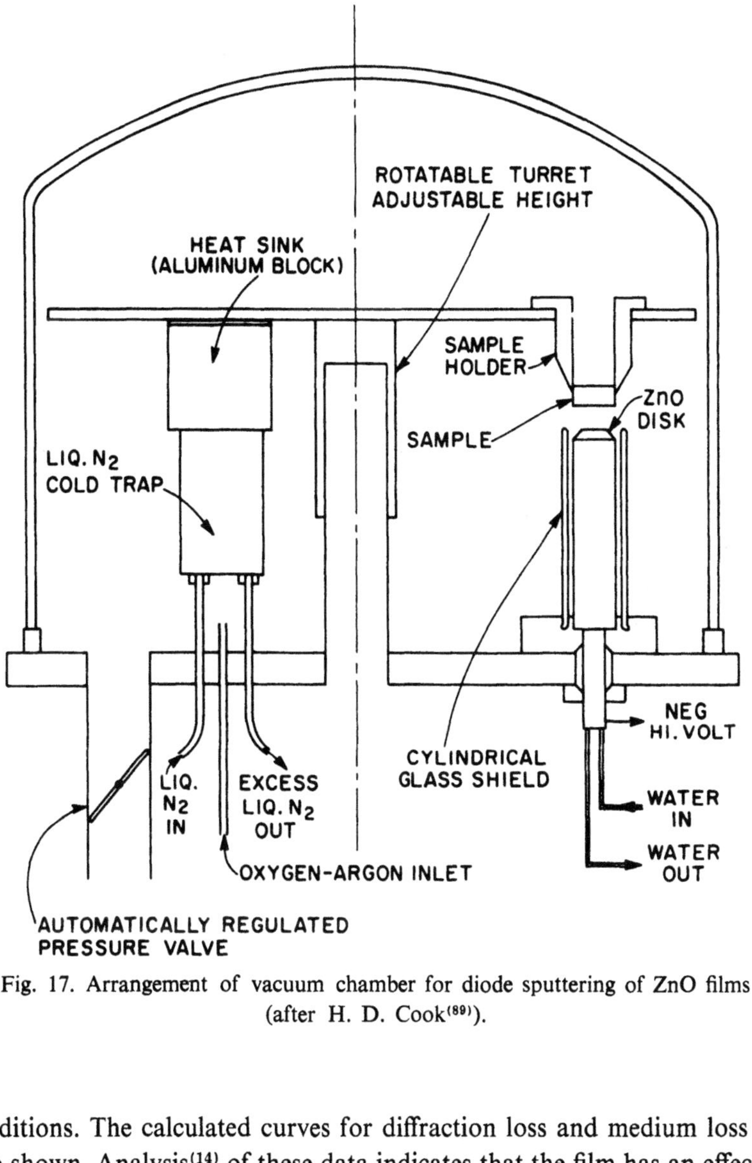

Fig. 17. Arrangement of vacuum chamber for diode sputtering of ZnO films (after H. D. Cook[89]).

conditions. The calculated curves for diffraction loss and medium loss are also shown. Analysis[14] of these data indicates that the film has an effective coupling factor of 0.24 ± 0.02.

Diode sputtered ZnO films have been made to perform well as fundamental mode thickness-dilatational transducers, even with films thinner than 2 μm. Figure 19 shows experimental data obtained by Denburg and Vannatta[90] on a ZnO film 1.8 μ thick. The experimental points are fitted by a theoretical loss curve calculated for the combination of metal layers, transducer film, and acoustic medium indicated in the figure. The values

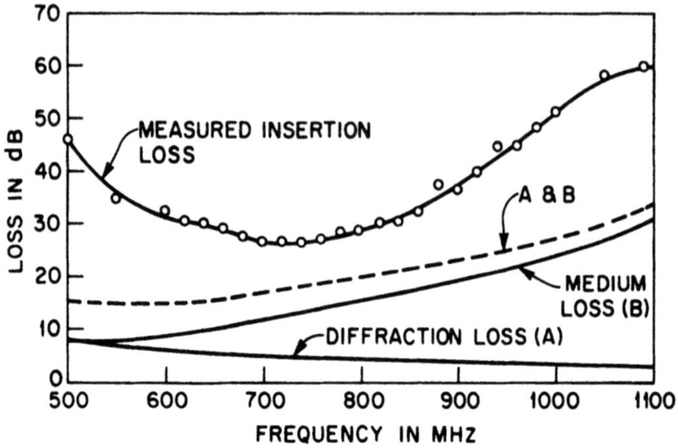

Fig. 18. Insertion loss *vs* frequency for a 3.2-μm-thick ZnO dilatational film prepared by diode sputtering. Curves for calculated diffraction loss and calculated medium loss are also shown.

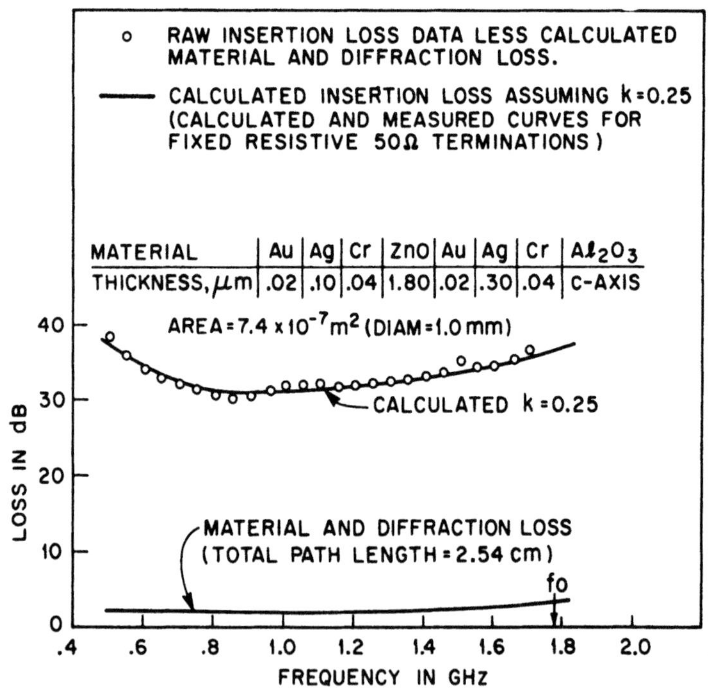

Fig. 19. Insertion loss *vs* frequency for a 1.8-μm-thick ZnO dilatational film prepared by diode sputtering. The sequence of metal electrode layers, bonding layers, and the transducer layer is indicated in the figure (after Denburg and Vannatta[90]).

of the densities and elastic constants used for the layers were the low-frequency bulk values; the value of coupling factor used for the ZnO film was 0.25, which is 88% of the bulk value. The wide flat transmission band obtained with this structure is particularly noteworthy and is a consequence of the high coupling factor of the ZnO film, the acoustic transmission properties of the metal layers, and the fixed resistive termination conditions used in the measurement.

Another useful sputtering technique is triode sputtering. In triode sputtering the processes of ionizing the gas and of bombarding the target material are largely separated by the use of a hot filament which emits electrons used to ionize the sputtering gas. The injection of electrons into the gas enables the discharge to be maintained at lower pressures. Because the sputtering process can be carried out at lower gas pressures for which the mean free path is several centimeters, the condensing vapor of sputtered material can have greater directionality. In contrast, the pressures required for diode sputtering result in mean free paths of the order of one millimeter. Since the source–substrate separation in sputtering systems is usually a few centimeters, the pressures required for diode sputtering result in a situation similar to indirect evaporation where there is no directionality left in the condensing vapor.

The basic arrangement used[79] for triode sputtering of ZnO films is shown in Fig. 20. A potential of +50 V is applied to the anode used to accelerate the electrons. The target is given a negative potential of −1500 V to attract positive ions from the plasma. The substrate on which the ZnO film is to be deposited is left floating at the potential of the plasma.

Both diode and triode sputtering techniques have produced transducers with dilatational-mode coupling factors in the range 0.24 ± 0.01, this high value resulting to a large extent from the tendency of ZnO films to grow with the *c*-axis approximately normal to the plane of the film. In spite of this tendency, it is possible to sputter ZnO films for shear transducers.[79] The technique of inclining the substrate surface relative to the vapor beam has not been as successful for ZnO as for CdS because the degree of control was not satisfactory. However, in contrast to the behavior of CdS films, ZnO films have shown the capability of growing with the *c*-axis in the plane of the film. Experiments by Foster[79] have indicated that the presence of organic material on the substrate surface enhances the probability of this type of film growth and that an organic compound such as xylene can be deliberately introduced into the vacuum system in order to promote the growth of such films. With this technique, coupling factors of 0.18, or 69% of the single-crystal value, have been obtained.

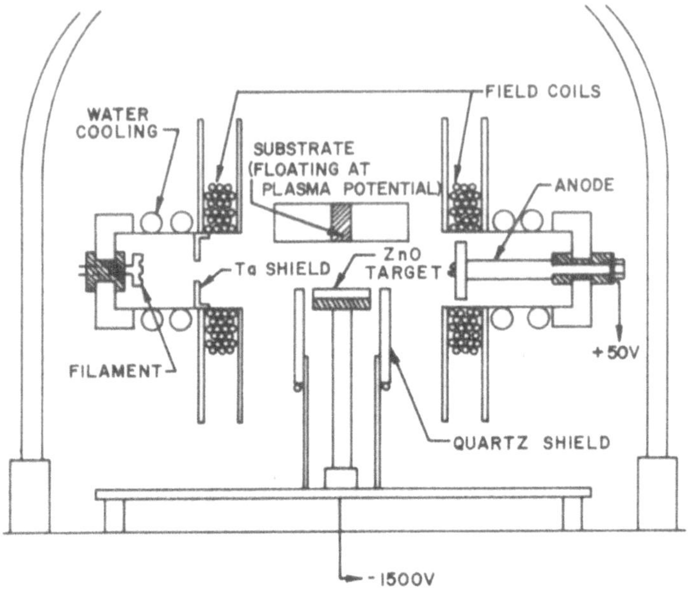

Fig. 20. Arrangement of vacuum chamber for use in triode sputtering of ZnO films (after Foster).

3.4. Other Compounds of Potential Interest for Film Tranducers

3.4.1. Aluminum Nitride (AlN)

CdS and ZnO are presently the most widely used compounds for film transducers; however the search for better materials is being pursued in a number of laboratories. AlN is a recently studied material that shows promise for film transducer applications. As yet, the constants for this material have not been fully determined; hence Table II provides only a listing for the thickness-dilatational mode.

The fabrication and performance of AlN piezoelectric films have been investigated by Wauk and Winslow.[91] They generated dilatational-mode films by a reactive evaporation technique. In this technique aluminum was evaporated and reacted with either nitrogen or ammonia gas to form AlN on metallic film substrates at rates up to 0.2 μm/min. The substrates were held at temperatures of 300 to 1200°C and films that were generated were oriented with the *c*-axis normal to the plane of the film. Films were obtained with discrimination against shear waves in excess of 60 dB.

Since AlN has (*6mm*) point group symmetry, films of this material might be expected to behave in much the same way as CdS and ZnO films.

The outstanding attractive property of this material is the high acoustic wave velocity, which in the thickness-extensional mode corresponds to a frequency constant of 5200 MHz · μm. This, along with a *K* of 8.5 makes AlN particularly attractive for transducers in the GHz frequency range. Wauk and Winslow presented data on AlN transducers operating at frequencies up to 5 GHz in the fundamental, but unfortunately their experiments were not of a type that enabled them to make quantitative determinations of the extensional-mode coupling factor. Nevertheless the results they obtained indicate that AlN is a potentially useful high-frequency transducer material and that further study of this material and related nitride compositions is warranted.

3.4.2. Films of Ferroelectric Materials

The high coupling factors available from single-crystal ferroelectric materials suggest rather naturally that these materials might be a means of achieving piezoelectric films with higher coupling factors than those presently available from CdS and ZnO. There is no known theoretical reason to believe that films a few microns thick cannot exhibit ferroelectric behavior. Polycrystalline ceramics with grain sizes significantly less than 2 μm in diameter are readily prepared; while the small grain size strongly influences the ferroelectric behavior of the material, it does not prevent it.[92,93] Furthermore, it should be remembered that CdS and ZnO have been estimated[94] to have at room temperature internal polarizations of 2.8 and 5.8 μCoul/cm^2, respectively. As Berlincourt has pointed out in Chapter 2, these materials are not ordinarily classified as ferroelectrics because they cannot be switched before dielectric breakdown occurs.

Sputtering conventional lead zirconate–lead titanate or sodium–potassium niobate ferroelectric ceramics is not a promising approach because these materials have such large dielectric constants that one usually encounters electrical impedance problems for transducers well within the limits of the bonded plate technology. Single-crystal ferroelectric materials with low dielectric constants and high coupling factors, like $Ba_2NaNb_5O_{15}$, appear to be much more attractive for exploration as materials for sputtered film transducers. In connection with the idea of sputtering single-crystal ferroelectrics, it should be noted that Foster[95] has reported some success in achieving piezoelectric films of $LiNbO_3$. These were *c*-axis films generated with a triode sputtering technique. The maximum coupling factor was 0.08 or about 47% of the single-crystal value. The best films were obtained with initial substrate temperatures in the range of 325 to 380°C. To be truly competitive with ZnO films for transducer applications, $LiNbO_3$ films (or films of other ferroelectric materials) would have to possess not only a high degree of preferential alignment of the grains but also be formable with orientations that combine high coupling factors and comparatively low dielectric constants with pure modes of vibration.

4. CONCLUDING REMARKS

The review material presented in this chapter describes the art of high-frequency piezoelectric transducers as developed up to 1970. The discussion has concentrated on the use of thin plates or films of piezoelectric materials as transducers for generating elastic body wave motions. In the later half of the 1960's, surface wave motions at frequencies extending into the GHz range came to be of considerable scientific and some technological interest. The discussion of surface wave transducer techniques and materials has not been included in this chapter primarily because this is a relatively new area of activity and one that is in a rapidly changing state of development.

In the decade of the 60's, the performance capabilities of transducer materials and techniques were greatly improved. This review has attempted to indicate that the improved capabilities have resulted from the development of both new materials and new techniques. The most important of these new materials have been the synthetically-grown piezoelectric and ferroelectric single crystals that combine high coupling factors with low dielectric constants. The most important new techniques are the sputter and vapor deposited thin-film transducer techniques that make possible the fabrication of fundamental-mode transducers operating at frequencies in excess of 1 GHz.

In regard to the performance capabilities of devices using piezoelectric transducer materials and techniques and their position relative to other competitive device techniques, the following observations seem appropriate. Above 100 MHz film transducer techniques involving CdS and ZnO become useful and above 500 MHz hold a dominant position. High-frequency ultrasonic delay lines using bonded-plate high coupling factor ceramic or single-crystal transducers hold a virtually unchallenged position for future applications as stores of serial digital data operating at bit rates in excess of 100 Mbits/sec. Acoustooptic devices using piezoelectric transducers will certainly find some applications; but since light steering and light modulation are functions that can be carried out by electrooptic devices, these are in competition with acoustooptic devices. Acoustooptic techniques have demonstrable superiority in many instances. Both materials and techniques presently exist to achieve low-loss wide-band ultrasonic and acoustooptic devices at frequencies up to 2 GHz.

REFERENCES

1. E. K. Sittig and H. D. Cook, "A method for preparing and bonding ultrasonic transducers used in high-frequency digital delay lines," *Proc. IEEE* **56**, 1375–1376 (1968).
2. A. H. Meitzler, "Advances in the use of ferroelectric materials as high-frequency piezoelectric transducers," paper presented at IEEE–CUA Symposium on Applications of Ferroelectrics, Washington, D.C., October 10–11, 1968.
3. D. Maydan, "Acoustooptical pulse modulators," paper presented at 1969 Conference on Laser Engineering and Applications, Washington, D.C., May 26–28, 1969.
4. D. A. Pinnow, "A solid-state acoustooptic light deflector," paper presented at 1969 Conference on Laser Engineering and Applications, Washington, D.C., May, 1969.
5. W. P. Mason, *Electromechanical Transducers and Wave Filters*, D. Van Nostrand Co., Inc., New York (1942, 1948), 2nd Ed., pp. 399–404.
6. J. E. May, Jr., "Characteristics of delay lines using quartz and barium titanate transducers," *J. Acous. Soc. Am.* **26**, 347–355 (1954).
7. H. J. McSkimin, "Transducer design for ultrasonic delay lines," *J. Acous. Soc. Am.* **27**, 302–309 (1955).
8. W. F. Konig, L. B. Lambert, and D. L. Schilling, "The bandwidth, insertion loss, and reflection coefficient of ultrasonic delay lines for backing materials and finite thickness bonds," *IRE International Convention Record* **9**, pt. 6, 285–295 (1961).
9. M. Onoe and A. H. Meitzler, "Effect of high electromechanical coupling on characteristics of piezoelectric transducers," Paper N44, Fourth International Congress on Acoustics, August 21–28, 1962.
10. C. F. Brockelsby, J. S. Palfreeman, and R. W. Gibson, *Ultrasonic Delay Lines*, Iliffe Book Ltd., London (1963).
11. E. K. Sittig, "Transmission parameters of thickness-driven piezoelectric transducers arranged in multilayer configurations," *IEEE Trans. on Sonics and Ultrasonics* **SU-14**, 167–174 (1967).
12. A. J. Bahr and Ian N. Court, "Determination of the electromechanical coupling coefficient of thin-film cadmium sulfide," *J. Appl. Phys.* **39**, 2863–2868 (1968).
13. E. K. Sittig, "Effects of bonding and electrode layers on the transmission parameters of piezoelectric transducers used in ultrasonic delay lines," *IEEE Trans. on Sonics and Ultrasonics* **SU-16**, 2–10 (1969).
14. A. H. Meitzler and E. K. Sittig, "Characterization of piezoelectric transducers used in ultrasonic devices operating above 0.1 GHz," *J. Appl. Phys.*, **40**, 4341–4352 (1969).
15. E. K. Sittig, "High-speed ultrasonic digital delay line design: a restatement of some basic considerations," *Proc. IEEE* **56**, 1194–1202 (1968).

16. N. F. Foster, G. A. Coquin, G. A. Rozgonyi, and F. A. Vannatta, "Cadmium sulfide and zinc oxide thin-film transducers," *IEEE Trans. on Sonics and Ultrasonics* **SU-15**, 28–41 (1968).
17. D. Berlincourt, "Delay line transducer materials," 1967 *IEEE International Convention Record*, Part 11, 61–68 (1967).
18. A. W. Warner and A. H. Meitzler, "Performance of bonded single-crystal $LiNbO_3$ and $LiGaO_2$ as ultrasonic transducers operating above 100 MHz," *Proc. IEEE* **56**, 1376–1377 (1968).
19. E. K. Sittig, A. W. Warner and H. D. Cook, "Bonded piezoelectric transducers for frequencies beyond 100 MHz," *Ultrasonics* **7**, 108–112 (1969).
20. J. de Klerk and E. F. Kelley, "Coherent phonon genration in the gigacycle range via insulating cadmium sulfide films," *Appl. Phys. Letters* **5** (1964).
21. N. F. Foster, "Ultrahigh frequency cadmium sulfide transducers," *IEEE Trans. on Sonics and Ultrasonics* **SU-11**, 63–68 (1964).
22. N. F. Foster, "Piezoelectric and piezoresistive films," in *Handbook of Thin Film Technology*, McGraw-Hill, New York (1970), Chap. 15.
23. A. R. Hutson, J. H. McFee, and D. L. White, "Ultrasonic amplification in CdS," *Phys. Rev. Letters* **7**, 237–239 (1961).
24. D. L. White, "Amplification of ultrasonic waves in piezoelectric semiconductors," *J. Appl. Phys.* **33**, 2547–2554 (1962).
25. D. L. White, "Depeltion layer transducer—a new high-frequency ultrasonic transducer," *1961 IRE International Convention Record*, pt. 6, **9**, 304–309 (1961).
26. D. L. White, in *Physical Acoustics* (W. P. Mason, ed.), Vol. I, Part B, Academic Press Inc., New York (1964), pp. 321–352.
27. N. F. Foster, "Diffusion layer ultrasonic transducers," *J. Appl. Phys.* **34**, 990–991 (1963).
28. F. S. Hickernell and D. E. Allen, "A GaAs diffusion layer transducer delay line," *Proc. IEEE* **53**, 1735 (1965).
29. N. Chubachi, H. Aoki, T. Seki, M. Wada, and Y. Kikuchi, *CdSe diffusion layer transducer*, Acoustical Society of Japan (May 1966).
30. Y. Kikuchi, N. Chubachi, and H. Sasaki, *Zinc oxide diffusion layer transducer*, Acoustical Society of Japan (November 1966).
31. F. S. Hickernell, "Diffusion layer transduction in semiconducting zinc oxide," Paper G3 presented at IEEE Ultrasonics Symposium, Vancouver, Canada, October 4–6, 1967.
32. F. S. Hickernell, "Piezoelectric semiconductor acoustic delay lines," *IEEE Trans. on Microwave Theory and Techniques* **MTT-17**, 957–963 (1969).
33. R. A. Laudise, E. D. Kolb, and A. J. Caporaso, "Properties of lithium-doped hydrothermally-grown single crystals of zinc oxide, *J. Am. Ceramics Soc.* **48**, 342 (1965).
34. H. Jaffe and D. Berlincourt, "Piezoelectric transducer materials, *Proc. IEEE* **53**, 1372–1386 (1965).
35. C. Solbrig, "Piezoelectric measurements on zinc oxide crystals," *Z. Phys.* **184**, 293–298 (1965).
36. D. F. Crisler, J. J. Cupal, and A. R. Moore, "Dielectric piezoelectric and electromechanical coupling constants of zinc-oxide crystals," *Proc. IEEE* **56**, 225–226 (1968).

37. R. T. Smith, "Temperature dependence of the electromechanical constants of Li-doped ZnO," paper presented at Acous. Soc. Am. 77th Meeting, Philadelphia, Penn., April 8–11, 1969.
38. J. T. Krause and W. R. Northover, U.S. Patent No. 3,413,187, "Glass bonding medium for ultrasonic devices," Nov. 26, 1968.
39. F. I. Federov, *Theory of Elastic Waves in Crystals*, Plenum Press, New York (1968). See Chap. 3.
40. R. B. Wilson, "Precision polishing of thin single-crystal layers," *J. Sci. Instruments* **44**, 159 (1967).
41. G. A. Bennett and R. B. Wilson, "Precision polishing technique for optics and microwave acoustics," *J. Sci. Instruments* **43**, 669–670 (1966).
42. H. J. McSkimin, in *Physical Acoustics* (W. P. Mason, ed.), Vol. I, Part A, Academic Press Inc., New York (1964), pp. 271–334. (See especially p. 318.)
43. H. J. McSkimin, "Measurement of ultrasonic wave velocities for solids in the frequency range 100 to 500 MHz," *J. Acous. Soc. Am.* **34**, 404–409 (1962).
44. J. Lamb, M. Redwood, and Z. Shteinshleifer, "Absorption of compressional waves in solids from 100 to 1000 MHz, *Phys. Rev. Letters* **3**, 28–29 (1959).
45. E. G. Spencer, R. T. Denton, T. B. Bateman, W. B. Snow, and L. G. Van Uitert, "Microwave elastic properties of nonmagnetic garnets," *J. Appl. Phys.* **34**, 3059–3060 (1963).
46. J. T. Krause, "Gold–indium bond for measurement of ultrasonic properties in solids at high temperatures," *J. Appl. Phys.* **39**, 5334–5335 (1968).
47. A. H. Meitzler and A. H. Fitch, "The acoustoelastic effect in vitreous silica, Pyrex, and T-40 glass," *J. Appl. Phys.* **40**, 1614–1621 (1969).
48. D. L. White, "β-quartz as a high-temperature piezoelectric material," *J. Acous. Soc. Am.* **31**, 311–314 (1959).
49. H. M. Matthews and R. C. LeCraw, "Acoustic wave rotation by magnon–phonon interaction," *Phys. Rev. Letters* **8**, 397–399 (1963).
50. J. P. Remeika and A. A. Ballman, "Flux growth, Czochralski growth, and hydrothermal synthesis of lithium metagallate single crystals," *Appl. Phys. Letters* **7**, 180–181 (1964).
51. A. W. Warner, "New piezoelectric materials," in *Proceedings of the 19th Annual Symposium on Frequency Control*, April 1965, pp. 5–21.
52. A. A. Ballman, "The growth and properties of piezoelectric bismuth germanium oxide $Bi_{12}GeO_{20}$," *International J. Crystal Growth* **1**, 37–40 (1967).
53. M. Onoe, A. W. Warner, and A. A. Ballman, "Elastic and piezoelectric characteristics of bismuth germanium oxide $Bi_{12}GeO_{20}$," *IEEE Trans. on Sonics and Ultrasonics* **SU-14**, 165–167 (1967).
54. E. J. Spencer, P. V. Lenzo, and A. A. Ballman, "Ultrasonic properties of bismuth germanium oxide," *Appl. Phys. Letters* **9**, 201–291 (1966).
55. S. Haussühl, "Piezoelectric and electric behavior of lithium iodate" (in German), *Phys. Stat. Sol.* **29**, K159–161 (1968).
56. A. W. Warner, J. G. Bergman, Jr., D. A. Pinnow, and G. R. Crane, "Piezoelectric and photoelastic properties of lithium iodate," to be published, *J. Acous. Soc. Am.* (1970).
57. E. G. Spencer, P. V. Lenzo, and A. A. Ballman, "Dielectric material for electro-optic, elastooptic, and ultrasonic device applications," *Proc. IEEE* **55**, 2074–2108 (1967).

58. A. A. Ballman, "Growth of piezoelectric and ferroelectric materials by the Czochralski technique," *J. Am. Ceramic Soc.* **48**, 112 (1965).
59a. K. Nassau, H. J. Levinstein, and G. M. Loiacono, "Ferroelectric lithium niobate. 1. Growth, domain structure, dislocations, and etching," *J. Phys. Chem. Solids* **27**, 983–988 (1966).
59b. K. Nassau, H. J. Levinstein, and G. M. Loiacono, "Ferroelectric lithium niobate. 2. Preparation of single-domain crystals," *J. Phys. Chem. Solids* **27**, 989–996 (1966).
59c. S. C. Abrahams, J. M. Reddy, and J. L. Bernstein, "Ferroelectric lithium niobate. 3. Single-crystal X-ray diffraction study at 24°C," *J. Phys. Chem. Solids* **27**, 997–1012 (1966).
59d. S. C. Abrahams, W. C. Hamilton, and J. M. Reddy, "Ferroelectric lithium niobate. 4. Single-crystal neutron diffraction study at 24°C," *J. Phys. Chem. Solids* **27**, 1013–1018 (1966).
59e. S. C. Abrahams, H. J. Levinstein, and J. M. Reddy, "Ferroelectric lithium niobate. 5. Polycrystal X-ray diffraction study between 24 and 1200°C," *J. Phys. Chem. Solids* **27**, 1019–1026 (1966).
60. D. B. Fraser and A. W. Warner, "Lithium niobate: a high-temperature piezoelectric transducer material," *J. Appl. Physics* **37**, 3853–3854 (1966).
61. A. W. Warner, M. Onoe, and G. A. Coquin, "Determination of elastic and piezoelectric constants for crystal in class [3*m*]," *J. Acous. Soc. Am.* **42**, 1223–1231 (1967).
62. B. T. Matthias and J. P. Remeika, "Ferroelectricity in the ilmenite structure," *Phys. Rev.* **76**, 1886–1887 (1949).
63. H. J. Levinstein, A. A. Ballman, and C. D. Capio, "Domain structure and Curie temperatures of single-crystal lithium tantalate," *J. Appl. Phys.* **37**, 4585 (1966).
64. A. W. Warner and A. A. Ballman, "Low-temperature coefficient of frequency in a lithium tantalate resonator," *Proc. IEEE* **55**, 450 (1967).
65. J. J. Rubin, L. G. Van Uitert, and H. J. Levinstein, "The growth of single-crystal niobates for electrooptic and nonlinear applications," *Int. J. Crys. Growth* **1**, 315 (1967).
66. L. J. Van Uitert, J. J. Rubin, and W. A. Bonner, "Growth of $Ba_2NaNb_5O_{15}$ single crystals for optical applications," *IEEE Trans. on Quantum Electronics* **QE-4**, 622–627 (1968).
67. J. E. Geusic, H. J. Levinstein, J. J. Rubin, S. Singh, and L. G. Van Uitert, "The nonlinear optical properties of $Ba_2NaNb_5O_{15}$," *Appl. Phys. Letters* **11**, 269–271 (1967). An error correction for the above paper, *Appl. Phys. Letters* **12**, 224 (1968).
68. A. W. Warner, G. A. Coquin, A. H. Meitzler, and J. L. Fink, "Piezoelectric properties of $Ba_2NaNb_5O_{15}$," *Appl. Phys. Letters* **14**, 34–35 (1969).
69. A. W. Warner, G. A. Coquin, and J. L. Fink, "Elastic and piezoelectric constants of $Ba_2NaNb_5O_{15}$," *J. Appl. Phys.* **40**, 4353–4356 (1969).
70. L. Egerton and D. M. Dillon, "Piezoelectric and dielectric properties of ceramics in the system potassium–sodium niobate," *J. Amer. Ceramic Soc.* **42**, 438–442 (1959).
71. L. Egerton and C. A. Bieling, "Isostatically hot-pressed sodium–potassium niobate transducer material for ultrasonic devices," *Am. Ceram. Soc. Bull.* **47**, 1151–1156 (1968).
72. D. Berlincourt, "Delay line transducer materials," 1967 *IEEE International Convention Record*, Pt. 11, 61–68 (1967).
73. R. E. Dean, unpublished memorandum.

74. J. S. Jones, "VHF ultrasonic delay line bonding," to be published in *IEEE Trans. on Sonics and Ultrasonics.*
75. D. R. Herriott, J. S. Jones, T. R. Meeker, and K. Reznicek, U.S. Patent No. 3,453, 166 "Method and apparatus for bonding transducer element," July 1, 1969.
76. E. K. Sittig, unpublished data.
77. D. L. Arenberg, U.S. Patent No. 2,754,238, "Method of bonding and article thereby formed," July 10, 1956.
78. D. Beecham, "Sputter machining of piezoelectric transducers," *J. Appl. Phys.* **40**, 4357–4361 (1969).
79. N. F. Foster, "Crystallographic orientation of zinc oxide films deposited by triode sputtering," *J. Vacuum Science and Technology* **6**, 111–114 (1969).
80. N. F. Foster and A. H. Meitzler, "Insertion loss and coupling factors in thin-film transducers," *J. Appl. Phys.* **39**, 4460–4461 (1968).
81. N. F. Foster, Cadmium sulfide evaporated-layer transducers," *Proc. IEEE* **53**, 1400–1405 (1965).
82. J. de Klerk, "Thin-film piezoelectric transducers used as generators and detectors of microwave phonons, with some attenuation measurements in SiO_2," *J. Appl. Phys.* **37**, 4522–4528 (1966).
83. J. de Klerk and E. F. Kelley, "Vapor deposited thin-film piezoelectric transducers, *Rev. Sci. Instr.* **36**, 506 (1965).
84. R. Weber, "Electron bombardment technique for deposition of CdS film transducers," *Rev. Sci. Instr.* **37**, 955–956 (1966).
85. N. F. Foster, "Structure of CdS evaporated films in relation to their use as ultrasonic transducers," *J. Appl. Phys.* **38**, 149 (1967).
86. R. M. Malbon, D. J. Walsh, and D. K. Winslow, "Zinc oxide film microwave acoustic transducers," *Appl. Phys. Letters* **10**, 9–10 (1967).
87. S. Wanuga, T. A. Midford, and J. P. Dietz, "Zinc oxide film transducers," paper presented at the IEEE Ultrasonics Symposium, Boston, Mass., December, 1965.
88. G. A. Rozgonyi and W. J. Polito, "Preparation of ZnO thin films by sputtering of the compound in oxygen–argon," *Appl. Phys. Letters* **8**, 220–221 (1966).
89. H. D. Cook, unpublished work.
90. D. L. Denburg and F. A. Vannatta, "Wide-band high-coupling ZnO transducers for microwave delay lines and optical modulators," paper presented at the 1969 IEEE Ultrasonics Symposium, St. Louis, Missouri, September, 1969.
91. M. T. Wauk and D. K. Winslow, "Vacuum deposition of AlN acoustic transducers," *Appl. Phys. Letters* **13**, 286–288 (1968).
92. C. E. Land and P. D. Thacher, "Ferroelectric ceramic electrooptic materials and devices," *Proc. IEEE* **57**, 751–768 (1969).
93. J. R. Maldonado and A. H. Meitzler, "Ferroelectric ceramic light gates operated in a voltage-controlled mode," *IEEE Trans. Electron Devices, ED-17,* 148–157 (1970).
94. Tomoya Ogawa, "Estimation of the spontaneous polarization of hexagonal ZnS, CdS and ZnO crystals," *J. Phys. Soc. Japan* **25**, 1126–1128 (1968).
95. N. F. Foster, "The deposition and piezoelectric characteristics of sputtered lithium niobate films," *J. Appl. Phys.* **40**, 420–421 (1969).
96. M. Onoe, H. F. Tiersten, and A. H. Meitzler, "Shift in location of resonant frequencies caused by large electromechanical coupling in thickness-mode resonators, *J. Acous. Soc. Am.* **35**, 36–42 (1967).

97. D. A. Berlincourt, D. R. Curran, and H. Jaffe, in *Physical Acoustics* (W. P. Mason, ed.) Vol. I, Academic Press Inc., New York (1964), Part A, pp. 169–270.
98. D. B. Fraser, J. T. Krause, and A. H. Meitzler, "Physical limitations on the performance of vitreous silica in high-frequency ultrasonic and acoustooptical devices," *Appl. Phys. Letters* **11**, 308–310 (1967).
99. J. T. Krause, unpublished data.
100. D. E. Chapin, "Frequency and temperature dependence of shear wave attenuation in Bausch and Lomb T-40 glass," *IEEE Trans. on Sonics and Ultrasonics* **SU-15**, 178–181 (1968).
101. D. B. Fraser, unpublished data.
102. W. P. Mason, *Physical Acoustics and Properties of Solids*, D. Van Nostrand Co., Inc. New York (1958), p. 17.
103. E. K. Sittig, unpublished data.
104. J. de Klerk, "Multilayer thin film piezoelectric transducers," *IEEE Trans. Sonics and Ultrasonics* **SU-13**, 99–103 (1966).
105. E. G. Spencer, R. T. Denton, T. B. Bateman, W. B. Snow, and L. G. Van Uitert, "Microwave elastic properties of nonmagnetic garnets," *J. Appl. Phys.* **34**, 3059–3060 (1963).
106. D. A. Pinnow, L. G. Van Uitert, A. W. Warner, and W. A. Bonner, "Lead molybdate: a melt-grown crystal with a high figure of merit for acoustooptical device applications," *Appl. Phys. Letters* **15**, 83–86 (1969).
107. A. H. Fitch, unpublished data.

INDEX

GPSR Compliance

The European Union's (EU) General Product Safety Regulation (GPSR) is a set of rules that requires consumer products to be safe and our obligations to ensure this.

If you have any concerns about our products, you can contact us on ProductSafety@springernature.com

In case Publisher is established outside the EU, the EU authorized representative is:

Springer Nature Customer Service Center GmbH
Europaplatz 3
69115 Heidelberg, Germany

Batch number: 10620163

Printed by Printforce, the Netherlands